Possible Worlds

Possible Worlds

J.B.S. Haldane
with a new introduction by
Carl A. Price

Routledge
Taylor & Francis Group

LONDON AND NEW YORK

Originally published in 1927 by Chatto & Windus

Published 2002 by Transaction Publishers

Published 2017 by Routledge
2 Park Square, Milton Park, Abingdon, Oxon OX14 4RN
605 Third Avenue, New York, NY 10017

Routledge is an imprint of the Taylor & Francis Group, an informa business

Library of Congress Catalog Number: 00-056797

Library of Congress Cataloging-in-Publication Data

Haldane, J.B.S. (John Burdon Sanderson). 1892-1964.
   [Possible worlds and other essays]
     Possible worlds / by J.B.S. Haldane; with a new introduction by
  Carl A. Price.
       p. cm.
   Originally published: Possible worlds and other essays. London: Chatto &
  Windus, 1927.
   ISBN 0-7658-0715-7 (pbk. : alk. paper)
   1. Science. 2. Medicine. 3. Religion and science. I. Title.
Q171.H154   2000
500—dc21   00-056797

ISBN 13: 978-0-7658-0715-1 (pbk)

TABLE OF CONTENTS

TRANSACTION INTRODUCTION

J. B. S. HALDANE'S POSSIBLE WORLDS: LOOKING AHEAD AT SCIENCE IN THE 1920S AND LOOKING BACK TODAY

JOHN Burdon Sanderson Haldane (1892–1964) was a giant among men. I know the appellation is overworked, but he deserves it. During the 1920s, '30s, and '40s, Haldane made major contributions to neurobiology, genetics, population biology, and evolution, particularly in providing a sound model for genetic change as a mechanism for evolution. He was at once comfortable in mathematics, chemistry, the emerging field of biochemistry, microbiology, and animal physiology. What's more, he was trained as a classicist as an undergraduate at Oxford.

Haldane was British and came from a noble family—an uncle, Viscount Richard Burdon Haldane, served in His Majesty's government during World War I. Haldane himself served as an officer (cf. the essay *Meroz,* p. 241) and was severely wounded, once in Flanders and once in Mesopotamia. His personal courage, which attracted public attention, was demonstrated by his eagerness to serve as guinea pig in physiological experiments, as, for example, in a depressurized chamber that mimicked the crest of Mount Everest (cf. the essay *On being one's own Rabbit,* p. 107).

Haldane stridently expressed his views about politics, economics, psychology, religion...whatever. Was there anything on which he didn't voice an opinion? Malcolm Muggeridge wrote of Haldane, "He was enormously energetic, opinionated, erudite, aggressive and, at heart, kindly and generous." And again, "With his colleagues he tended to be irascible and suspicious; his professional career was a long series of rows in which, in his own estimation, he was invariably in the right."

But there were other able scientists of his time, other brave men and women, and (I'm sure) equally opinionated writers, but we are not necessarily reprinting their books. Haldane was distinctive as someone who promoted science and his ideas about science to ordinary people with a persuasive, lively, even humorous journalistic style—a role that Irving Louis Horowitz identifies as that of *public intellectual*. Within a very few years, he wrote an astonishing number of essays, collected in *Possible Worlds*, that were originally published in journals as varied as the *Daily Mail,* the *Manchester Guardian, Harper's Magazine,* the *Atlantic Monthly,* and the *New Republic.* And his writings had a powerful influence on public opinion.

Among scientists today, only Stephen Jay Gould comes even close to Haldane in his skill in reaching the nonscientific reader.

Possible Worlds was first published in 1927. That date has resonance for me; it was the year I was born.

So, while the period of 1920 to 1927 does not seem like ancient history to me, I am struck by how profoundly our world has changed since then. Part of the change, of course, is in science itself, but Haldane's essays bring home vividly just how much technology has changed the lives and outlook of ordinary people.

Let us skim through some of the essays for Haldane's interpretations of science; his views on society, art, and religion; and his predictions for what science will bring in the future, which is our present.

Science in the 1920s

Some Dates

In this essay Haldane writes of geology (p. 16):

> ...the earth has lasted for at least a thousand million [= billion] years in a condition not very unlike the present...

We now know the earth was formed four to six billion years ago. He qualifies his one billion with "at least," so his time scale is not all that far off. Since, however, we now have evidence that the earth's atmosphere changed from anaerobic to aerobic only about one billion years ago, we can be confident that the earth is in fact greatly changed from its original condition. (The gradual accumulation of free oxygen in the atmosphere, from zero to the present-day 20 percent, was a product of photosynthesis by cyanobacteria.)

Vitamins

Haldane refers (p. 51) to the nutritional value of proteins in the diet of humans:

> A certain minimum of proteins, which must be of the right quality, is needed for repairs of the tissues.

The specific amino acids essential to human nutrition had not been identified in 1927. Moreover, the measurement and identification of amino acids in foods was extraordinarily difficult, requiring many kilograms of material. This was changed overnight around 1950 with the invention of *paper chromatography* (cf. my comments in *An Omitted Method* at the end of this introduction).

Also in *Vitamins* we find (p. 52) that

> ...at least five rather complex organic bodies are needed...

Depending on how one counts, there are more nearly a dozen vitamins, and—since they are all small molecules, rather than polymers—we would not think of them today as particularly "complex." Although no vitamin had yet been characterized when the essay was written, Albert Szent-Györgyi would soon describe the isolation of vitamin C. (In his first submission Szent-Györgyi knew only that vitamin C was a sugar, actually a sugar alcohol. Since sugars are routinely named with the suffix *ose*—e.g., sucr*ose* or fruct*ose*—he proposed to call it *Godnose*. Who said that scientists don't have a sense of humor?)

Oxygen Want

Haldane wrote of an area of physiology in which he had achieved public attention by performing experiments on himself. He realized that the problems of supplying oxygen to patients was largely technical: how to design a system for the administration of oxygen while allowing the patient to exhale (and discharge carbon dioxide) normally. He writes (p. 73):

> Oxygen has a great future in medicine, and could probably halve the death-rate in pneumonia. But as generally administered it has little more therapeutic value than extreme unction, and is much more expensive.

Water Poisoning and Salt Poisoning

In this essay we have another dramatic example of the simplifications that chemistry and physics have brought to the clinical laboratory. Haldane is writing of the importance of determining the contents of certain metal ions in blood (pp. 79-80):

> A competent biochemist will not err more than one per cent. in his estimation of the calcium in ten cubic centimetres of blood, but the analysis requires some hours, and competent biochemists are rare.

Hours spent in a single analysis of calcium in blood was transformed to a rapid and idiot-proof procedure around 1960 by the development of *flame photometry*. Now a machine aspirates a few drops of diluted blood into a special kind of torch, and within seconds the content of calcium, magnesium, and similar metals is determined from each metal's characteristic emission spectrum.

The Fight with Tuberculosis

Tuberculosis was perhaps the most feared scourge in Europe and North America through the first half of the twentieth century. The storied alpine villages of Switzerland had been famous in Haldane's time, not as ski resorts, but for their tuberculosis sanitaria. In *The Fight with Tuberculosis* he writes on how to protect one's family from infection (p. 98):

> The greatest single channel of [tuberculosis] infection is milk from tuberculous cows drunk in infancy or rarely childhood. But the vast majority even of well-to-do parents do not take the trouble to obtain Grade A or Grade A certified milk...

Now this was written twenty-five years after the death of Louis Pasteur, and how many of us today can remember when milk was *not* pasteurized? Haldane frequently expresses concern for microbial pathogens, how to avoid them and how to treat them. Alexander Fleming would discover penicillin a few years after publication of *Possible Worlds,* but the first effective treatment for tuberculosis would not be available until 1945, when Selman Waksman and his team would discover streptomycin. (Waksman also coined the term *antibiotic*.)

Haldane also expresses his concern for tuberculosis in *Occupational Mortality* (p. 197):

> The most dangerous of all occupations, with a death-rate almost two and a half time the average, is that of barman....[T]hey are more than twice as likely to perish of consumption and other lung diseases....These diseases are mainly due to the overcrowding and under-ventilation of their places of work...so dark and close as to form ideal breeding-grounds for the tubercle bacillus."

Should Scientific Research be Rewarded?

Haldane asks *Should Scientific Research be Rewarded?* He notes that successful writers, artists, musicians can expect comfortable incomes, but (p. 176):

> A scientist...is generally glad if he gets £1000 per year.

So, he asks, should people who make important discoveries be given monetary prizes as a token of their contribution to society? Although Haldane finds that the "justice of the contention seems clear," he finds a serious reservation, noting (p. 177) that

> It was a long time before Faraday's discovery of electro-magnetic induction could be applied to the manufacture of a practical dynamo....
>
> The greatest difficulty of a scheme of rewards during life is to be found in the impossibility of estimating the importance of a discovery until the discoverer is dead or too old to enjoy the money.

In an aside that seems incredible today, Haldane finds (p. 176) that

> Very little can be made by taking out patents and…medical
> etiquette rightly forbids the patenting of a new medicine or a
> new surgical instrument.

I never knew that there was a time when drugs or antibiotics could not be patented? Where would the pharmaceutical industry be today if that tradition had been sustained?

Some Enemies of Science

Although considered on the fringe, we have today vigorous defenders of *animal rights*. In Haldane's time the catchword was *antivivisectionists*. Haldane found them an enemy of science and lamented their effect on medical training (p. 250):

> Nor may the doctor…acquire surgical skill by an operation on
> an anaesthetized animal, even in a licensed laboratory.

I knew about antivivisectionists because my family strongly opposed experimentation on animals. (They were also vegetarians, but as a teenager I rebelled.) I did not realize that experimentation on mammals was so sharply restricted in Great Britain that surgeons could not practice with animal models.

Haldane's Views on Society, Art, Religion, the Poor

As I noted in the opening paragraphs, Haldane voiced his opinions on all manner of topics. Some of his essays focus specifically on social or philosophical issues (e.g., *When I am Dead,* p. 204), but more often his views appear in passages ostensibly devoted to science.

Art

In *Darwinism Today*, for example, he inserts put-downs for the writings of Edith Sitwell and the sculptures of Jacob Epstein into a discussion of evolution (p. 43):

> And it is difficult not to compare some of the fantastic animals of the declining periods of a race with the work of Miss Sitwell, or the clumsy but impressive with that of Epstein.

Incidentally, I note that Haldane uniformly refers to women as *Miss* or *Mrs.*, but men by their last names only. Since he does not otherwise display more than casual prejudice towards women, perhaps it was simply a convention of the time.

Religion

This essay on *Thomas Henry Huxley* appears to have been written in 1925 on the centenary of Huxley's birth. Haldane greatly admired Huxley, who was an eloquent proponent of evolution and defender of Darwinism in the nineteenth century (and who, we learn, coined the word *agnosticism*). In praise of Huxley, Haldane writes (p. 132):

> If the majority of educated Englishmen to-day reject the miraculous element in religion and the infallibility of the Bible, the result is due to Huxley more than to any other man, and in

particular to his extraordinary fairness of argument and moderation of language.

I expect these perceptions describe educated Englishmen today; I'm not so sure that a majority of educated Americans today "reject the miraculous element in religion."

In *Science and Theology as Art Forms* Haldane speaks of impediments that Christian churches throw in the path of science, with the decline of the West as a likely consequence (p. 234):

> The Christian churches are preventing people from thinking about life. In the interests of theological orthodoxy Christian children may not be taught about evolution....If any Asiatic people begins as a whole to think biologically before those of European origin do so, it will dominate the world...

Society and Politics

In his essay on Huxley, Haldane shifts to a favorite theme: that social interactions, economics, and politics are governed by scientific principles (p. 134):

> [Humanity] is subject to the same laws as those which govern the animals from which it has arisen. Until the mass of our people are convinced of this fact and ready to act upon it, Huxley's work will not be done.

Haldane was a Marxist, which was not that unusual among his peers (e.g., Desmond Bernal, Lancelot Hogben, and Joseph Needham). Later on, in the 1930s, he joined the Communist Party. I expect that his political views, at least in part, drove his conviction that human behavior was subject to scientific principles. The Communists claimed, for example, that the Soviet state had created a new *Soviet man*, who was qualitatively different from the peasants, working class, or (God forbid!) bourgeoisie who had come before; their novel virtues were a product of their socialist environment. The contrary notion, that the qualities of human beings were fixed in their genes, was heresy, so that by the 1950s Mendelian genetics was viewed as counterrevolutionary in the Soviet Union. Haldane was a Marxist, but he was also a rationalist. When the Soviets espoused the hypotheses of Trophim Lysenko, a plant breeder who claimed that environment could influence heredity, Haldane resigned from the party!

The Scientist's Ego

In *William Bateson,* an essay on an early proponent of Mendel's theory of genetics, Haldane admires three characteristics of Bateson (pp. 135-136). (The second of these, describing an irascible loner, may reflect Haldane's estimate of his own personality!)

>...when Mendel's paper in the Brunn Society's journal was discovered in 1900, Bateson had already hit upon the atomic theory of heredity, which goes by the name of Mendelism. It was characteristic of him that no hint of this fact is to be found in his published work...
>
>And his public attacks on the Darwinian theory [of evolution by natural selection] were so phrased as inevitably to lead to the most heated argument, and even to the extraordinary misrepresentation that he disbelieved in evolution.

...the last years of his life were largely given over to the investigation of exceptions to [Mendel's law]; and we owe to him more than to any other one man the demonstration not only that they are valid over a vast range of material, but that they occasionally break down.

I very much agree that the readiness to question the very principles on which one's reputation rests is indeed unusual among scientists. I have another example: After Hans Krebs convinced the world of the generality of the tricarboxylic acid cycle (*Krebs cycle*), the reactions at the core of respiration in aerobic organisms, he directed his associates to determine mechanisms that would account for apparent *exceptions* to it! A notable result was Hans Kornberg's elucidation of the glyoxylate cycle.

Population Limits and Family Size

As noted above, Haldane was a rationalist and a socialist. It is surprising, from today's perspectives, that he would not have recognized the dangers of overpopulation. Quite the opposite, he repeatedly promoted possibilities for harnessing natural resources to meet people's needs. He insists that, contrary to Malthus, technology will resolve all problems of an increasing population. In *Vitamins* (p. 50):

The upper limit to human numbers is not set by any facts of nature, but by human ignorance and inadaptability.

In *Daedalus, or Science and the Future* (which is not reproduced in *Possible Worlds,* but is drawn upon heavily in *The Future of Biology*), he cheerfully predicts the disappearance of agriculture:

> ...mankind will be completely urbanized. Personally I do not regret the probable disappearance of the agricultural labourer in favour of the factory worker, who seems to me a higher type of person...

And in *Should Scientific Research be Rewarded?* (p. 180):

> [Researchers should have an income] large enough to allow them to bring up a family of five children....[To] content themselves with one or two children [is] a questionable advantage...since scientific ability is strongly inherited.

Now, we take it as given today that the ideal number of children is about two, but many of my cohorts, the generation that begat baby boomers, had four, five, and more children. And they didn't have to be Catholic! Haldane, however, urges the "upper classes" to have more children, and frequently pushes the envelope in his assumptions about what human characteristics are inherited. Perhaps I'm being too hard on Haldane; infant mortality in those days, even among the upper classes, would have pushed two children per family below the rate of replacement.

Nationality, Ethnicity, and Race

While Haldane appears to be exceptionally enlightened on matters of race and ethnicity, he also displays some blatant prejudices about non-British nationalities, ethnicities, and races. A passage in *Nationality and Research* (p. 154) displays views that may have been widely held at the time but, by today's perspectives, reek of Nazism:

> For the Jews, just as they are partly responsible for one of the worst features of our civilization, the control of industry by financiers more interested in profit than service, have shown in other fields the most single-minded devotion to pure thought.

And Americans are second-rate scientists (p. 156):

> The United States produce a colossal volume of scientific work, of very unequal merit. Where endowment can assure results, they lead the world. Their astronomical observations form the bulk of international output, though their interpretation often comes from England, Germany, or Holland. In the studies of animal breeding and nutrition, the methods largely devised in Cambridge and London are being developed on a colossal scale. [Thomas Hunt] Morgan's work on inheritance in New York has involved the counting of over twenty million small flies....
>
> In spite of these facts and the undoubted genius of many Americans, I am inclined to think that in pure (though not perhaps in applied) science America produces less than either Britain or Germany.

So, who produces the best science? (p. 159):

> The largest actual output of scientific work comes from Britain and Germany....[I]t would be idle to deny the splendour of Germany's achievements at the present moment, more especially in such fields as organic chemistry and mathematics, which the Germans have made peculiarly their own.
>
> If Great Britain leads the world in many branches of science, it is, I think, largely through two causes, the autonomy of our universities, and the lack of nationalism in our scientific

thought. A university governing itself may be a little deaf to the claims of working-class education or the equality of the sexes, but it is more likely to appoint the best man to a post than is one governed by business men or politicians.

In fairness, we should remember that, seventy-five years ago, there were only limited opportunities for a first-class education in the sciences in the United States, and many of America's brightest went to Europe and Britain for their graduate degrees.

Lest I leave the impression of Haldane as a hopeless snob, his final comment on the subject (p. 161):

Scientific ability is not the perquisite of any one race but it can only show itself under conditions when thought is free...

Poison Gas

In an introductory note to *Eugenics and Social Reform*, Haldane laments the perversion of science. He writes (p. 190):

Perhaps the greatest tragedy of our age is the misapplication of science....[T]he principal result of many increases in human power and knowledge has been either an improvement in methods of destroying human life and property, or an accentuation of economic inequality.

Hearing these phrases today, we think of the prostitution of science and technology in warfare, of biological warfare, of poison gas. But Haldane is criticizing the "confused thinking of 'advanced' politicians" that rejected the use of poison gas by the Allies in World War I! (p. 190):

I refer to mental processes such as that which led to our forgoing the use of 'mustard gas,' the most humane weapon ever
invented, since of the casualties it caused, 2.6 percent died, and
1/4 percent were permanently incapacitated. No one at Washington even suggested abandoning H.E. [high explosives] and
shrapnel, which kill or maim about half their casualties.

As a neurobiologist, it is perhaps just as well that he
could not foresee the uses of nerve gas, of the development of chemical agents that kill instantly on contact.

Eugenics

In the early part of the twentieth century, the eugenics movement attracted wide interest. It purported
to improve quality in human populations by encouraging reproduction by superior people and discouraging reproduction (or promoting sterilization) of the unfit.
In *Eugenics and Social Reform* Haldane addresses the
scientific problems underlying eugenics (pp. 190-191):

The relevant facts [concerning heredity] fall into two classes,
first those which relate to hereditary abnormalities or tendencies to disease, and secondly those bearing on the inheritance
of intelligence and the different birth-rate in different social
classes…[with respect to inherited disease] we should first try
to impress on [affected people] their duty to restrict their families, and to see that they have the means to do so.
    We must first examine the question how far heredity rather
than environment is responsible for the mental differences between the children of different social classes. The question
cannot be answered on *a priori* grounds.

I note that he avoids the term *birth control*. Again
(pp. 195-196):

With regard to childbirth, rich women need more exercise, poor women more education.

To sum up, the rational programme for a eugenist is as follows: Teach voluntary eugenics by all means; but if you desire to check the increase of any population or section of a population, either massacre it or force upon it the greatest practicable amount of liberty, education, and wealth. Civilization stands in real danger from the over-production of 'undermen.' But if it perishes from this cause it will be because its governing class cared more for wealth than for justice.

Haldane was rightly appalled that (p. 190),

...the growing science of heredity is being used in [Britain] to support the political opinions of the extreme right, and in America [referring to forced sterilization] by some of the most ferocious enemies of human liberty.

But his writings also betray the notion that class was (is?) a fact of life in Great Britain. As a socialist *and* a member of the upper class, Haldane was sincerely solicitous of the working class and the poor; he frequently cites measures that should be taken on their behalf. But today's liberals would rip out their hard drives before admitting to the class consciousness displayed in Haldane's most casual remarks.

Despite his strongly held views on social issues, Haldane is not dogmatic. In *Science and Politics* he writes (p. 189):

And until politics are a branch of science we shall do well to regard political and social reforms as experiments rather than short cuts to the millennium.

I should add that he firmly believed that politics *would* become a scientific discipline.

Science as Philosophy

In *Science and Theology as Art Forms* Haldane places the scientific method into a Kantian perspective (p. 228):

> ...the experience of the past makes it clear that many of our most cherished scientific theories contain so much falsehood as to deserve the title of myths. Their claims to belief are that they contradict fewer known facts than their predecessors, and that they are of practical use. But there is one very significant feature of the most fully developed scientific theories. They tell us nothing whatever about the inner nature [*das Ding an sicht*] of the units with which they deal.

Looking into the Crystal Ball

In an essay on *The Future of Biology* Haldane predicts the future of science. As a widely read scientist—and an exceptionally bright person—his thoughts are especially interesting. While eager to speculate, Haldane first disavows the role of prophet (p. 139):

> ...one can be quite sure that the future will make any detailed predictions look rather silly.

But was he on target? Or would he be utterly shocked by what has developed? The answer, I think, is "yes" to both alternatives. He was sometimes on target; he was sometimes in left field; and he sometimes missed the boat.

On Animal Behavior

As we shall see repeatedly, Haldane has high hopes for discovering laws with predictive value in psychology and animal behavior, which Haldane calls *animal psychology* (p. 140):

> To-morrow it looks as if we should be overhearing the conversation of bees, and the day after to-morrow joining in it. We may be able to tell our bees that there is a tin of treacle for them if they will fertilize those apple trees five minutes' fly to the south-east....Talking with bees will be a tough job, but easier than a voyage to another planet.

While there was a piece recently on computer-assisted translation of dog-talk, it is evident that (at least unmanned) interplanetary travel has beaten out conversation with bees.

On Ecology (p. 141):

> ...we are constantly finding that some hitherto unexpected...factor, such as the acidity of the soil or the presence of some single parasite...will make a whole new fauna and flora appear....But as we find the key chemical or key organism in a given association, we may be able vastly to increase the utility to man of forests, lakes, and even the sea.

Except perhaps for the construction of dams or the eradication of wolves, I know of no single factor that effected such transformations of ecosystems. What I do find striking, as noted above, is Haldane's enthusiasm for exploiting natural resources. Clearly, *green* had not yet been invented.

Economics, Sociology, and Politics

With a hubris barely imaginable today, Haldane expects that the future will see theories of economics, sociology, and politics with predictive value, analogous to those in the physical sciences (p. 142):

> ...from the quantitative study of animal and plant associations some laws...are emerging; laws of which much that we know of human history and economics only constitute special...cases....When we can see human history and sociology against a background of such simpler phenomena, it is hard to doubt that we shall understand ourselves and one another more clearly.

On Evolution

As we have noted before, Haldane was very much concerned with evolution. Looking far into the future, he imagined, perhaps playfully, that one could achieve a certain precision in determining when evolutionary lines diverged (p. 142):

> ...our ideal is to establish a family tree of plants and animals: to be able to say definitely, let us say, that the latest common ancestor of both man and dog was... living...in what is now the North Atlantic 51,400,000 years ago, under the shade of the latest common ancestor of the palm and beech trees, while the last common ancestor of the dog and bear lived only 5,200,000 years back. We are still thousands of years from this ideal...

One could quibble over Haldane's calendar: his placement of the man-dog and palm-beech ancestors at 51.4 million years was doubtless based on the fossil record, and is not terribly far off the mark; 5.2 million years for the dog-bear ancestor is much too recent. But the fascinating aspect of his prediction is that it will take "thousands of years" to generate such a model. Haldane and his contemporaries could not imagine how the sequencing of proteins, and then of nucleic acids, together with estimates of mutation rates, would lead to models that predict the intervals separating groups of organisms, and, furthermore, that these predictions could be verified by radioactive dating.

On Genetics

Of special significance are Haldane's predictions in genetics (p. 144):

> We take any animal or plant, and...should be able to answer the following questions...:
> 1. What inheritable variations or mutations arise in it and how are they inherited?
> 2. Why do they arise?
> 3. Do they show any sign of being...of advantage to their possessor?
> 4. Would natural selection acting on such...account for evolution...?
>
> ...the question how to alter a single gene without interfering with the others becomes serious....The two most hopeful methods seem to be to find chemical substances which will attack one gene and not another; and to focus ultra-violet rays on a fraction of a chromosome...
>
> Until we can force mutations in some such way...we can only alter the hereditary composition of ourselves, plant, and animals by combining in one organism genes present in several, and so getting their combined effect.

Yes, we now know how to answer Haldane's four questions. Moreover, we know how to alter a single gene, and we know how to introduce highly specific mutations. But not by the procedures that Haldane found "most hopeful."

While these essays were written, Haldane held the position as Reader in Biochemistry at Cambridge and, shortly after publication of *Possible Worlds*, was appointed as part-time geneticist at what is now the John Innes Centre. Few people at that time were as well situated to imagine the future of genetics and what we now call genetic engineering. Yet, he did not, could not, imagine how genetics would transcend the techniques on which genetics then relied: breeding of experimental organisms and plotting the inheritance of characters in family trees. Haldane's essay was written fifteen years before George Beadle proposed that one gene encodes one enzyme (well, not quite), twenty years before Oswald Avery, C. M. McLeod, and M. McCarty presented the first evidence that genes were made of DNA, and twenty-five years before James Watson and Francis Crick were to described the structure of DNA as a double helix. In an event that would have been impossible to predict, genetics was utterly transformed. While plant and animal breeding continues largely as it was practiced in Haldane's time, genetics today has become a far more precise and incredibly more powerful and productive study of genes at the molecular level.

On Heredity and Class

As we have seen, Haldane was especially interested in human heredity, and he had great expectations for the applications of genetics to the human condition. While curiously circumspect about the heritability of certain diseases, he is very positive on the heritability of intellectual abilities and behavior, characters that we regard today as more likely to be influenced by the child's environment and by *multiple* genes, which can greatly complicate the patterns of inheritance. Unlike writers of today, and in seeming conflict with his socialist ideal of a classless society, Haldane routinely referred to *classes* of mankind with the seeming conviction that they were determined genetically (pp. 144-145).

> A great deal may thus be done with man...[but] many of the deeds done in America in the name of eugenics are about as much justified by science as were the proceedings of the Inquisition by the gospels.
>
> The first thing to do in the study of human heredity is to find characters which vary sharply so as to divide mankind definitely into classes.
>
> When a baby arrived we should have a physical examination and a blood analysis done on him, and say something like this: "He has got iso-agglutinin B and tyrosinase inhibitor J from his father, so it's twenty-to-one that he will get the main gene that determined his father's mathematical powers; but he's got Q4 from his mother...so it looks as if her father's inability to keep away from alcohol would crop up in him again...

On Photobiology

Photobiology has become a powerful tool in our understanding of all manner of phenomena from photosynthesis to the mutagenic effects of ultraviolet light. Haldane's expectations (p. 147) were more limited and have not come to pass:

> We treat skin tuberculosis with ultraviolet light. Can we increase the curative effect without increasing the danger of severe sunburn? ...The application of rays will gradually be taken out of the doctor's hands. He will write out a prescription, and we will go round to the radiologist's shop next door to the chemist's and ask for the prescribed treatment in his back-parlour.

On Neurobiology

Writing on another of his consuming interests, neurobiology, Haldane predicts great strides in understanding how the global functioning of the brain affects human behavior (p. 147):

> We have by now gone most of the way in the localization of function [in the brain], for although a given area of the brain is always concerned in moving the hand, yet a given point in it may cause different movements at different times; just as any one telephonist in an exchange can only ring up certain subscribers....Here we are still in the graph and table stage, but probably only about ten years off a fairly comprehensive theory of how the different parts of the nervous system act on one another. This will at once react on psychology, and more slowly on normal life and practical medicine....And until psychology is a science, scientific method cannot be applied in politics.

When it comes to psychology, my impression is that seventy-five years later we are *still* in the "graph and table stage."

On Biochemistry

Haldane's knowledge of biochemistry was exceptional among biologists of his time, but his expectations, in retrospect, were modest (p. 148):

> In chemical physiology we are after two rather different things. The first is to trace out the chemical processes in the cells, the nature, origin, and destiny of each substance in them. The second, which is much easier, is to trace the effect on cell life of various substances...

The first part, which we would call *metabolic maps*, is pretty complete. The second ceased to be of great concern to biochemists around mid-century, once it became evident that few chemicals taken off the shelf offer great specificity in the site of their action. What Haldane could not have expected was the discovery that specific proteins, or cascades of proteins, link external and internal stimuli to the control of expression of specific genes. Scientists of the 1920s had only a crude notion of the structure of proteins and, as noted above, no clue to the structure of genes.

On Hormones

Although the biochemistry of hormones was in its infancy, Haldane had high hopes for hormone therapy (p. 150):

A number of [hormones] have been obtained in a fairly concentrated form....Only two have been obtained entirely pure...
      When we have these substances available in the pure state
we ought to be able to deal with many departures from the
normal sexual life, ranging from gross perversion to a woman's
inability to suckle her children....We shall also probably be
able, if we desire, to stave off the sudden ending of a woman's
sexual life between the ages of forty and fifty.

The widespread benefits of replacement doses of
thyroxin, of various female hormones, and of growth
hormone support Haldane's prediction. In targeting
"gross perversion," could he be referring to homosexuality? And I guess that modern women were
never told that menopause causes the "ending of a
woman's sexual life."

On Cell and Tissue Culture

Haldane predicted (p. 151) that cell and tissue
culture would lead to tissue and organ replacement
in humans:

Cultures from individual cells from a chicken can be kept alive
in suitable media for twenty years, and as far as we know for
ever....Fifty years hence we shall probably know whether it is
worth seriously trying to obtain perpetual youth for man by
this method. A hundred years hence our great-grandchildren
may be seeing the first results of such attempts.

Until very recently such expectations would have
invited a firm rebuke from mammalian cell biologists: "Perhaps you didn't realize, my good man, that
cell differentiation in higher animals is totally irreversible!" That was before Dolly, the sheep. And now,

as the clock ticks towards the end of the millennium, the possibilities of regenerating a new heart, kidney, or liver from an individual's own stem cells are being seriously addressed; when achieved, this will confer in us a kind of "perpetual youth."

As a plant scientist, I have to gloat: for 150 years plant types have been comfortable with *totipotentiality,* the notion that individual cells are capable of regenerating replicas of the organism from which they came. In the 1960s, F. C. Steward demonstrated that whole carrot plants could be regenerated from root cells grown in tissue culture. Shortly afterwards, Georges Morel turned the orchid industry upside down by regenerating thousands of identical orchid plants from cell lines cultured from the stem of a single prize orchid. The procedure is now routine.

Haldane expected that the answers to "perpetual youth" would be known—yes or no—by 1975. Why was it delayed? Although the technology that produced Dolly from the nucleus of a fully differentiated mammary cell was not particularly sophisticated, I expect that the crucial experiments were delayed by the prevailing dogma in the field of mammalian cell biology.

The fruits of tissue culture, Haldane projected, would be harvested in industry as well as medicine (p. 152):

> We can now kill an animal and produce a fluid from inorganic constituents that will keep its heart or liver alive for a day or more....We could grow human embryos in such a solution, for

their connection with their mother seems to be purely chemical. We could cut our beefsteak from a tissue culture of muscle with no nervous system to make it waste food in doing work, and a supply of hormones to make it grow as fast as that of an embryo calf.

Haldane first promoted the possibility of *ectogenesis*, culturing mammals outside the womb, in *Daedalus, or Science and the Future,* an essay published in 1923. Aldous Huxley was obviously influenced by such thoughts in his writing of *Brave New World.* But while we do not grow babies in test tubes, nor have cattle ranches gone the way of the blacksmith, we *have* learned how to substitute dialysis for kidneys, maintain people on heart-lung machines, etc.

On Pharmacology

The discovery of new and better drugs, first among substances isolated from plants and microorganisms, then supplemented by synthetic organic chemistry, has unquestionably transformed our lives and protected us from innumerable diseases. Haldane is especially concerned with the management of pain (p. 152):

A few of the complicated substances made by plants have a striking effect on animals....If we had a drug that was as good a pain-killer as morphine, but one-tenth as poisonous and not a habit former, we could use it indiscriminately; and wipe out a good half of the physical pain in human life at one stroke.

This magic painkiller has not surfaced, but *pain management* has become much more sophisticated, with epidural nerve blocks and morphine-delivery devices that permit the patient to adjust his own dosage, usually with lower overall consumption.

An Omitted Method

Haldane expected wonderful results from a number of procedures that we should today dismiss as hair-brained. And he was acutely aware that the then current knowledge of biological compounds was extremely limited. For example (p. 227):

> ...no one has seen a toxin in the pure state.

But an extraordinarily powerful procedure was ignored by him and most of his colleagues; that was *chromatography*, a method developed around 1900 for resolving mixtures of plant pigments into pure compounds. While the general principles of chromatography are applicable to almost all classes of chemicals, for fifty years the method remained restricted to the separation of chlorophylls and carotenoids. Then, in the late 1940s, chromatography was extended to colorless compounds, resulting in a virtual explosion in biochemistry and chemistry. It did not require a crystal ball, but a visit to the lab down the hall. How many contemporaries of Haldane kicked themselves, saying, "Why didn't I think of that?"

Conclusions

So, if J. B. S. Haldane was the pompous, class-conscious, opinionated Big Professor that I may have conveyed in these notes, why should we read him? While exasperated at times in my reading of *Possible Worlds,* I found Haldane's clear, direct, and vivid prose to provide penetrating insights into an earlier society, and one that was mother to today's economy, politics, and science. That his ideas on class and race are revolting by today's standards, or that his political beliefs were naive, are not important in themselves; he spoke for many like-minded people, and that *is* important.

Haldane was truly exceptional in translating the science of his time into ideas that Everyman could grasp. His predictions of what science would achieve were on target more often than not, but his failed predictions are perhaps the most interesting of all, because they throw light on the truly novel ideas of the last seventy-five years.

Carl A. Price

J. B. S. HALDANE
(1892–1964)
A BRIEF CHRONOLOGY*

1892
John Burdon Sanderson Haldane was born
J. B. S. Haldane was from a Scottish family and son of John Scott Haldane, a physiologist at Oxford. His uncle, Viscount Richard Burdon Haldane, was Britain's Lord Chancellor during World War I

Attended Eton; studied classics and mathematics at Oxford

1911
Published his first scientific paper

1914–1918
Joined the Black Watch as a bombing officer during World War I, served in the Middle East and Flanders

1921–1937
Named *Reader in Biochemistry* at Oxford; applied mathematics to genetics

1923
Published *Daedalus, or Science and the Future,* which expanded on "A paper read to the Heretics," Cambridge, on Feburary 4, 1923

1925
Published *Callinicus: A Defence of Chemical Warfare*

1927
Published *Possible Worlds*

1927–1937
Appointed as *Head of Genetical Work* at the John Innes Horticultural Institution (now the *John Innes Centre*). Retaining his readership at Cambridge, Haldane served part-time at an annual salary of £400

1929
Published (with Julian Huxley) *Animal Biology*

1930
Published *Enzymes*

1931
Published *The Philosophical Basis of Biology*

1932
Published *The Causes of Evolution* and *The Inequality of Man and Other Essays.*

---

* Based on notes by Rosemary Harvey, former Archivist of the John Innes Institute.

1934   Published *Fact and Faith* and *Human Biology* and *Politics*

1937   Named *Weldon Professor of Biometry* at University College London
Published *My Friend Mr. Leakey* (a children's book)
Joined the Communist Party [date not certain, "mid-1930s"]

1938   Published *Heredity and Politics* and *The Marxist Philosophy and the Sciences*

1940   Published *Science in Peace and War*
Appointed chairman of the editorial board of the *Daily Worker* (through 1949)

1941   Published *New Paths in Genetics, Keeping Cool and Other Essays,* and *Science and Everyday Life*

1946   Published *A Banned Broadcast, and Other Essays*

1947   Published *Science Advances*

1949   Published *What is Life?*
Strongly rejected the Communist Party's endorsement of Lysenko's genetics

1951   Published *Everything Has a History*

1954   Published *The Biochemistry of Genetics*

1956   Resigned from the Communist Party

1957   Emigrated to India in protest over the seizure of the Suez Canel by the British and French

1968   Published posthumously *Science and Life: Essays of a Rationalist*

[uncertain date of publication]
On Being the Right Size and Other Essays

PREFACE

THE essays collected in this book have mostly, but not all, appeared in print. In Europe they have appeared in the *Rationalist Annual*, the *Bermondsey Book*, the *Nation*, the *Daily Mail*, the *World To-Day*, the *Manchester Guardian*, the *Graphic*, the *Weekly Dispatch*, *Discovery*, *Modern Science*, and the *Haagsche Maandblad*. In America they have been published by *Harper's Magazine*, the *Forum*, the *Century Magazine*, the *Atlantic Monthly*, and the *New Republic*. They have been written in the intervals of research work and teaching, to a large extent in railway trains. Many scientific workers believe that they should confine their publications to learned journals. I think, however, that the public has a right to know what is going on inside the laboratories, for some of which it pays. And it seems to me vitally important that the scientific point of view should be applied, so far as is possible, to politics and religion. In such spheres the scientific man cannot, of course, speak with the same authority as when he is describing the results of research ; and in so far as he is scientific he must try to suppress such of his own views as have no more scientific backing than those of the man in the street.

Some of these essays are on medical topics. As I do not hold a medical degree I can speak more freely than a qualified physician. But if a doctor cannot answer questions with regard to individual cases which he has not examined, an unqualified person is still less able to do so. I have rarely written on a medical subject without receiving letters from would-be patients. It is obvious that I cannot answer such communications.

The essays in the first part of this book deal mainly with matters of fact. Those which follow are more speculative. In scientific work the imagination must work in harness. But there is no reason why it should not play with the fruits of such work, and it is perhaps only by so doing that one can realize the possibilities which research work is opening up. In the past these results have always taken the public and the politicians completely by surprise. The present disturbed condition of humanity is largely the result of this unpreparedness. If the experience is not to be repeated on a still greater scale it is urgent that the average man should attempt to realize what is happening to-day in the laboratories.

ON SCALES

'LE silence éternel de ces espaces infinies m'effraie,' said Pascal, as he looked at the stars and between them, and his somewhat irrational terror has echoed down the centuries.

It is fashionable to find the distance of even the nearest fixed stars inconceivable, and to make no attempt to grapple with the number of atoms in one's thumbnail. And this habit of mind makes it quite unnecessarily hard for the plain man to understand the main results of modern science, many of which are quite straightforward, but happen to involve rather large numbers. For Pascal's attitude is neither scientific nor religious. ' I shall soon be above that fellow,' said Sir Thomas More, as he took his last look at the sun before his execution, and the modern astronomer views the sun as a rather small but quite fairly typical star in a particular cluster.

There is no reason to suppose that interstellar space is infinite. Very probably the whole of space is finite, and certainly the distances of all the visible heavenly bodies are within the range of the human mind. Infinity is the prerogative of mind rather than matter. We can reason about it, but we certainly cannot and do not observe it. As for the silence of interstellar space, one could not live in it, and hence could not discover whether it is silent or not. But if one were shut up in a steel box in it, like Jules Verne's travellers to the moon, one would probably hear fairly frequently (at least in the neighbourhood of a star) the sound made by a minute dust particle moving at enormous speed hitting one's abode.

The average man complains that he cannot imagine the eighteen billion miles which is the unit in modern astronomy when once we leave the solar system, and is called a parsec because the apparent parallax of a star at this distance is a second ; in other words the earth's orbit from a parsec away would subtend an angle of a second, or look as large as a halfpenny at six thousand yards' distance. Of course one cannot imagine a parsec. But one can think of it, and think of it clearly.

For every educated person learns a process which is really of extraordinary difficulty, and involves a stupendous change of scale. That process is map-reading. In ordinary life our practical unit is about a centimetre, or two-fifths of an inch. Rather few of the measurements of everyday life exceed this in accuracy. Now suppose we look at a map of the world on a globe measuring sixteen inches round the equator, we are using a model on a scale of one in a hundred million (10^{-8}) and the average man learns to understand its meaning and draw practical information from it. An Englishman hears that his son is going to New Zealand, and has only to look at the globe to see that his letters will take longer to arrive than those from his other son in Newfoundland. But although we are at home on this particular scale, of 1000 kilometres or about six hundred miles to a centimetre, as regards the earth, the average person has not yet grasped the fact that on the same scale the sun is a mile off and as large as a church.

Our grandchildren will have learnt to do the opposite mental trick, namely, to be familiar with models on a scale of a hundred million to one. On this scale the atoms of the common elements are represented as less than an inch across, and molecules of fairly complex organic substances are a foot or so long. The electrons in these atoms and the nuclei round which they are believed to circulate would still be too small to be visible, but we could mark out their orbits, just as in a map we can represent railway lines, though only by exaggerating their width. It is doubtful whether a much greater magnification would serve any real purpose. When we come to deal with the events inside the atom the attempt to represent them in space and time breaks down, or at any rate the properties of space and time in very small quantities are so unlike those of common-sense space and time, that models are of rather slight value. On the other hand models of chemical molecules deduced from X-ray analysis of crystals are most reliable guides, and are opening up a new era in chemistry.

Let us now take a second step in the opposite direction and try to construct a model such that in it the globe will be as much reduced as the earth had been in representing it as the globe. That is to say our model is to be on a scale of one in ten thousand million million (10^{-16}). This would really show us very little, for not only the earth, but its orbit round the sun would be invisibly small, and even the orbit of Neptune would be comfortably contained on a pin's head, which would also represent the size of the largest known star. Unfortunately however even on this scale the nearest fixed star would be four yards away, and only about a hundred would be within thirty yards. The Milky Way would be a good day's walk across. Light would creep much more slowly than a snail, but quicker than the growth of most plants !

A third step in the same direction would probably be legitimate. If we reduced our scale once more by a hundred million times, the Milky Way would be almost invisibly small, and the nearer spiral nebulae would be represented only a fraction of an inch away from it, while probably all the spiral nebulae visible with the best telescope could be represented within a radius of half a mile. It is not clear that we could repeat the operation a fourth time. For the extended theory of relativity seems to lead inevitably to the view that the universe is finite, and that progress in any direction would ultimately lead back to the starting-point. In fact an attempt to make a model on this scale would perhaps produce results as misleading as those obtained when, by Mercator's projection, we try to represent the surface of the earth on a single plane. On the fourth order model the volume of the whole of space might be as small as one hundred thousandth of a cubic millimetre, though this is a lower limit.

We have seen then that we can usefully think of models up to a hundred million times life-size, and down to a scale of about a million million million millionth. Beyond those limits space does not have the properties ascribed to it by common sense, and visual imagination does not help us. We are compelled to plunge into the mathematics of the quantum theory at the small end, of relativity at the big end. But long before that is necessary, people are frightened off the attempt to think, apparently by the word ' million.' This is because it is generally applied to large aggregates like a million pounds or a million years, which we cannot easily imagine, though as a matter of fact a quite ordinary room would hold a hundred million gold sovereigns, provided its floor did not give way. But we ought to get the million habit by remembering that we wash ourselves daily in a bath containing about ten million drops of water, and often walk ten million millimetres during the day.

It is a pity that outside India no opportunities are presented of seeing a million men and women, for crowds of this size only occur on Hindu religious pilgrimages, and very impressive they are. A crowd of three million may sometimes be seen at the Kumbh Mela, a twelve-yearly festival which, if I remember, will next be held at Allahabad in January 1930. I can cordially recommend attendance there to any one who cannot imagine a million. Incidentally I am informed that participation in it gets one off several million reincarnations.

In science we soon get accustomed to these large numbers. The astronomer switches over merrily enough from measuring stellar distances in kiloparsecs, which take light 3000 years to travel, to determining its wave-length correct to a fraction of an Angstrom unit, which is a hundred-millionth of a centimetre. And there is a certain thrill when the final result of a calculation which has involved hundreds of millions comes out at one or two, when up till the last moment it might apparently have been anything from a million to a millionth, and thus leads to a simple theory. I am thinking, for example, of Eddington's famous calculation as to why stars are no heavier (for none are known as much as a hundred times heavier than the sun). Starting from the data of physics he calculated the internal temperatures of the stars; and since radiation exerts a push on matter emitting, absorbing, or reflecting it, he was able to discover what proportion of the weight of a star of given mass was supported by its own radiation. This proportion is negligible for stars lighter than the sun, but increases to half in a star about five times the sun's weight, while any star much heavier would burst. Thus through a wilderness of millions we arrive at a rational explanation of why all stars have about the same weight.

Again Gorter and Grendel, and Fricke and Morse, have shown by quite independent methods that the oily film surrounding a red blood corpuscle is just two molecules thick. Gorter extracted the oil, and spread it out in a film on water, Fricke measured the electrostatic capacity of the corpuscles by putting blood in a very rapidly alternating electric field. Both used numbers including the five thousand million corpuscles in a cubic centimetre and the six hundred thousand million million million atoms in a gram of hydrogen, but the final answer was 'two' in the one case, and 'one or two' in the other. It is the success of such calculations that makes it impossible for a scientifically trained person to disbelieve in the numbers on which they are based.

SOME DATES

FIVE hundred years ago the human mind was limited to a tiny patch of space, and the universe must have seemed even smaller after Magellan's men had girdled the earth. The heavenly bodies were known to be distant, but it was not clear that celestial distances were so much greater than terrestrial. So Cassini's proof, in A.D. 1672, that the sun was nearly a hundred million miles away was at first too shocking a fact for the mind to accept. Only eighty-nine years ago Bessel measured the distance of one of the nearest fixed stars, 700,000 times greater than that of the sun, and to-day Hubble and other astronomers are estimating distances several million times those that staggered our great-grandfathers.

The range of our minds in time is also increasing, but the process has been slower, partly because time is harder to measure than space, and partly because the chronology of the Old Testament is more precise than its astronomy. So when it was admitted that the earth was older than the six or eight thousand years which the biblical record allowed, scientific men were at first very moderate in their estimates of geological time. Twenty-five years ago geologists and physicists would not admit that the earth could be more than twenty million years old, although the biologists were asking for hundreds of millions for the process of evolution.

In the last generation, however, evidence has accumu-

lated along at least five different lines which allow us
to measure the past with complete accuracy for nearly
four thousand years, and with tolerable exactitude for
over a thousand million. We may conveniently begin
with the nearer dates which fall within the range of
history. A generation ago the earliest date known
with any certainty was that of the first Olympic Games,
776 B.C. Even if the accuracy of the ages of the
patriarchs in the book of Genesis was accepted, the
length of time between the births of Jacob and David
was very uncertain ; and the dates fixed by Archbishop
Ussher were to that extent at least conjectural.

But there were certain records of eclipses on historic
occasions whose dates were known within a few years.
Now a total eclipse of the sun visible from any given
place is a very rare event ; indeed only five have been
visible in any part of the British Isles since A.D. 1433.
So if we know the place where the eclipse was total, and
the date within a century or so, we can calculate the
latter with great accuracy. Every one, for example,
has heard of Tweedledum and Tweedledee, whose
battle was interrupted by a monstrous crow as big as a
tar-barrel. The true story of these heroes is as follows :
King Alyattes of Lydia, father of the celebrated Croesus,
had been engaged for five years in a war with Cyaxares,
king of the Medes. In its sixth year, on May 28,
585 B.C., as we now know, a battle was interrupted by
a total eclipse of the sun. The kings not only stopped
the battle, but accepted mediation. One of the two
mediators was no less a person than Nebuchadnezzar,
who in the preceding year had destroyed Jerusalem
and led its people into captivity. Other eclipses

recorded by the Assyrians enable us to date their kings who were contemporary with the kings of Judah and Israel, and incidentally show us that Archbishop Ussher was forty-six years out in his chronology of that period. This is no discredit to the learned prelate, but is highly disgraceful to the publishers who continue to print bibles containing it, and the clergy who continue to use them. Whoever else may have been inspired, Archbishop Ussher was not, and we need not pay much attention to clergymen who protest their reverence for Scripture, and yet continue to use, or permit their flocks to use, bibles adorned with the conjectures of an Irish divine whose political talents were at least as marked as his intellectual.

Readers of Homer will remember that Odysseus' return to Ithaca was marked by an eclipse of the sun which portended the doom of Penelope's suitors. As early as A.D. 1612 the attempt was made to date the fall of Troy by this means. But it was only in 1925 that Dr. Schoch of Munich, using far more exact tables of the moon's motion, arrived at the startling result that in the year 1178 B.C. there actually was a total eclipse of the sun in or very near to Ithaca at 11.41 A.M. on April 10th. Since the track of an eclipse is only 120 miles broad at most and generally less, and Ithaca is only 15 miles long, the sun has probably not been totally eclipsed in Ithaca since Odysseus' time, or for thousands of years before. Now the most probable date for the siege of Troy was generally given at about 1200 B.C., so presumably both this date and Homer's story of the eclipse were approximately correct. One need not suppose that the suitors were actually killed

on the day of the eclipse; but for the hero's return and
the darkening of the sun to become connected in local
tradition, as they apparently were, they must have
occurred within a few years of one another.

The first date which is known with nearly complete
certainty is 1915 B.C. We have accurate tables of the
appearances and disappearances of the planets in the
reign of King Ammizaduga, tenth king of the first
dynasty, who reigned in the city of Babylon from 1922
to 1902 B.C. In the sixth year of his reign, for example,
we read on a cuneiform tablet recently discovered,
' In the month Arahsamnu on the 28th day Venus
disappeared in the west (*i.e.* as an evening star).   Three
days she tarried in heaven, and rose in the east on the
first day of Kislev.'   On the basis of such data as these,
Father Kugler, a German Jesuit, and Dr. Fotheringham
of Oxford, have been able to arrive at the only possible
system of dates which will fit the facts.

It is indeed fortunate that King Assurbanipal, who
reigned in Nineveh from 668 to 626 B.C. approximately,
was so addicted to astrology as to have copies made of
the observations of predecessors who, when he lived,
were already as remote as are King Penda of Mercia
or the Caliph Ali to-day.   The goodness of the agreement between dates found astronomically and those
derived from lists of Mesopotamian dynasties has
augmented the faith of historians in the latter, and by
their use Professor Langdon of Oxford has calculated
back to about 3357 B.C. as the date of the beginning of
the second dynasty in the city of Ur.   Unfortunately
the kings who are recorded as having lived before this
date are often alleged to have reigned for many cen-

turies. If we allow them lives of a reasonable length we arrive at a date for the great Mesopotamian flood somewhere between 5000 and 6000 B.C. This event, which is probably historical, though greatly exaggerated, will not be fully explained till Iraq and Armenia have been studied by competent geologists. At this time a good deal of Scandinavia was still covered by an ice sheet left over from the last glacial epoch, and the same was probably true of Armenia. Noah's flood may well have been due to an abnormal thaw, perhaps accompanied by the bursting out of a lake or lakes pent up behind a glacier or moraine.

For in Scandinavia and Canada the melting ice has left very exact records, which Baron de Geer and his pupils have investigated. The whole of Scandinavia, 12,000 years ago, was covered by ice. Then the covering of its southern tip began to melt, and each year the thaw water from it deposited a layer of mud. At any given spot a number of such layers may be found wherever a road or railway cutting or a pit allows the examination of the subsoil. The thick layers due to warm years which thawed much ice can easily be identified. As one travels northward each layer is gradually overlaid by fresh ones and finally disappears. As lately as 9000 years ago the site of Stockholm was still covered by ice, but now the icefields are restricted to high ground. The final 7000 years in de Geer's calculations were reached by the counting of annual layers of clay laid down in a lake. In Canada the northern ice-sheet probably reached the great lakes less than 20,000 years ago, though here the evidence is not so complete. De Geer's counting of the mud bands gives us an idea of the geological time scale. There were four ice ages during the Pleistocene period. The last of them was already waning 20,000 years ago, and as there were lengthy warm periods between them, the whole Pleistocene period must have lasted for some hundreds of thousands of years, perhaps the best part of a million. Similar bands, if they consist of mud laid down in annual floods, record the work of a great river in Burma in mid-tertiary times during about a million and a half years.

But the principal evidence for the geological time-scale is of a different kind. Uranium and thorium break down into a series of short-lived radio-active elements which end up as lead. If the rate of decay has always been the same as at present, half of any given mass of uranium is transformed in the course of about 4,600,000,000 years. The fixity of this rate may seem a large assumption. But it is justifiable for two reasons. Firstly no chemical or physical treatment has the slightest effect on it. Secondly the speed with which a-particles are shot out from radio-active atoms depends on their rate of decay. Now particles of radio-active matter in mica and other rocks are surrounded by definite spheres of discoloration where the a-particles from them have stopped. If the velocity and hence the range of these particles had altered during geological time these spheres would not be definite. Assuming then that the ' clocks ' have not slowed down or speeded up one can use them to calculate the age of the rocks in the following way. Many volcanic minerals contain uranium or thorium but very little lead. But there is always some lead ;

and the older the rock, as judged by ordinary geological standards, the more lead is present. From its quantity we can calculate how long the change has been going on. This gives us the following ages for various strata. (B.M. means before man. It does not matter what individual man we consider !) :—

Eocene (London Clay)   . 60 million B.M.
Carboniferous (British coal   260-300 million B.M.
    measures).
Upper Pre-Cambrian   . 560 million B.M.
Oldest known rock   .   . about 1500 million B.M.

These dates may be as much as ten per cent. out, but can hardly be a great deal more.

That is to say, 60,000,000 years ago our ancestors were mammals, probably not unlike lemurs, 300,000,000 years ago amphibians somewhat resembling newts or mud-puppies, and 500,000,000 years ago very primitive fish, combining some of the characters of sharks and lampreys. The origin of life on our planet was probably over a thousand million years ago, so that the record furnished by fossils only refers to half—perhaps much less than half of the time during which life has existed.

If all the lead in our planet is of radio-active origin, which is rather unlikely, it can hardly be more than eight thousand million years old. Astronomical evidence points to a somewhat smaller age.

As the earth goes round, the moon, and to a lesser extent the sun, raise tides in the sea. The energy used in raising them comes from the earth's rotation, hence they slow it down and lengthen the day. The moon thus acts as a brake on the earth, and by so doing is pushed onwards in its orbit, and moves further away. If we calculate backwards instead of forwards we find both the day and the month becoming shorter, until at a sufficiently early date they possessed the same length of about four hours, and the moon was so near to the earth as to be practically touching it. It is fairly clear that the moon is a portion of the earth thrown off as the result of excessive rotation, almost certainly before the earth's crust had solidified. Unfortunately the frictional effect of the tides depends on the detailed form of the sea's bed. At present the main retarding action takes place in the Bering Sea. At a geological epoch characterised by many shallow and partly land-locked seas tidal friction must have been greater than now, at other times less. So we can only say that the moon was born somewhere about four thousand million years ago, but the true figure might be as low as one thousand million, or as high as twenty thousand.

The birth of the moon was only one event in a greater catastrophe. Our sun, after a relatively brief period, probably a few thousand million years or less, of youthful exuberance as a giant star radiating energy at thousands of times its present rate, settled down as a respectable dwarf, which it now is, and has been throughout geological time.

For many thousands of millions of years it probably shone as a lonely star unaccompanied by planets. Then it appears to have passed near to another, probably heavier star, which raised tidal waves in it. The detached crests of these waves, or one of them, formed the planets, and it is fairly clear that the moon broke off from the earth within a few years of its formation. So the approximate dating of the moon's birth gives us that of the earth's. This is further confirmed by the eccentricity of Mercury's orbit, which is still far less circular than the earth's, but is gradually settling down towards circularity. It can be calculated that it has not been going round the sun for more than ten or less than one thousand million years. Various other lines of evidence converge to a date somewhere between 8,000,000,000 and 1,500,000,000 B.M. for the origin of the solar system. If science continues we shall arrive at the exact date in the following way. The relative motions of the various ' fixed ' stars will be determined, and on calculating backwards it will be found that one passed very near to our sun at a certain date in the remote past. The star in question must be very far away by now. It is a wise child that knows its own father, and we shall probably not know ours for thousands, perhaps hundreds of thousands of years.

It is possible to penetrate still further into the past and to arrive at a very rough date for the origin of the sun. But any such date depends on some particular hypothesis as to the origin of stellar energy, and there are several such hypotheses, leading to very different dates. On the other hand a number of independent arguments, based on well-ascertained facts, converge to the same date for the origin of the earth. There are, of course, respectable scientific theories, such as the planetesimal, which lead to different conclusions. The reason for rejecting such theories, and the detailed evidence for many of the dates here given, are to be found in such books as Jeffrey's *The Earth* (Cambridge

University Press). In a popular exposition it has been necessary to be dogmatic. If I have been so it is because I consider it unlikely that any of the figures I have given will be very seriously upset in the future.

In a few generations it is probable that these dates will meet with general acceptance and their meaning will gradually penetrate the human imagination. As the earth has lasted for at least a thousand million years in a condition not very unlike the present, it will probably continue habitable for a future period of at least the same order of magnitude, possibly for very much longer. An acceptance of such a future is bound to affect human thought. It will be realized that the things which seem to us most stable, such as human nature and the facts of geography, are really not only changeable but certain to change. On the other hand it will be realized that remarkably little change can occur within a lifetime.

Such a world-view leaves room for optimism in the most desperate circumstances, but yet reduces the probable effects of the vastest human efforts to the tiniest dimensions. As it is accepted, people will probably become more and more prone to devote themselves to their own affairs and those of their immediate neighbours. And when they turn their attention to greater things, they will perhaps be less occupied with institutions as ephemeral as nations. They will be more disposed to serve Man than England or America. A just law may outlive the state in which it was made, a scientific discovery the civilization which brought it forth.

And religion will inevitably alter its standpoint, even if some of its fundamental beliefs survive. On a planet more than a thousand million years old it is hard to believe—as do Christians, Jews, Mohammedans, and Buddhists—that the most important event has occurred within the last few thousand years, when it is clear that there were great civilizations before that event. It is equally difficult to doubt that many events as significant for humanity will occur in the future. In that immeasurable future the destiny of humanity dwarfs that of the individual. If our planet was created a few thousand years ago to end a few years or a few thousand years hence, it is conceivable that the main purpose to be worked out on it is the salvation and perfection of individual human beings. No religion which accepts geology can regard such a purpose as anything but subsidiary.

If we define religion as our attitude to the universe as a whole, the new time-scale will make us humbler as individuals, but prouder as a race. Our individual lives are the merest spangles of existence. The life of our ancestors goes back for a thousand million years. That of our descendants may last very much longer. And we cannot say with any certainty that it will not endure for ever.

ON BEING THE RIGHT SIZE

THE most obvious differences between different animals are differences of size, but for some reason the zoologists have paid singularly little attention to them. In a large textbook of zoology before me I find no indication that the eagle is larger than the sparrow, or the hippopotamus bigger than the hare, though some grudging admissions are made in the case of the mouse and the whale. But yet it is easy to show that a hare could not be as large as a hippopotamus, or a whale as small as a herring. For every type of animal there is a most convenient size, and a large change in size inevitably carries with it a change of form.

Let us take the most obvious of possible cases, and consider a giant man sixty feet high—about the height of Giant Pope and Giant Pagan in the illustrated *Pilgrim's Progress* of my childhood. These monsters were not only ten times as high as Christian, but ten times as wide and ten times as thick, so that their total weight was a thousand times his, or about eighty to ninety tons. Unfortunately the cross sections of their bones were only a hundred times those of Christian, so that every square inch of giant bone had to support ten times the weight borne by a square inch of human bone. As the human thigh-bone breaks under about ten times the human weight, Pope and Pagan would have broken their thighs every time they took a step.

This was doubtless why they were sitting down in the picture I remember. But it lessens one's respect for Christian and Jack the Giant Killer.

To turn to zoology, suppose that a gazelle, a graceful little creature with long thin legs, is to become large, it will break its bones unless it does one of two things. It may make its legs short and thick, like the rhinoceros, so that every pound of weight has still about the same area of bone to support it. Or it can compress its body and stretch out its legs obliquely to gain stability, like the giraffe. I mention these two beasts because they happen to belong to the same order as the gazelle, and both are quite successful mechanically, being remarkably fast runners.

Gravity, a mere nuisance to Christian, was a terror to Pope, Pagan, and Despair. To the mouse and any smaller animal it presents practically no dangers. You can drop a mouse down a thousand-yard mine shaft; and, on arriving at the bottom, it gets a slight shock and walks away, so long as the ground is fairly soft. A rat is killed, a man is broken, a horse splashes. For the resistance presented to movement by the air is proportional to the surface of the moving object. Divide an animal's length, breadth, and height each by ten; its weight is reduced to a thousandth, but its surface only to a hundredth. So the resistance to falling in the case of the small animal is relatively ten times greater than the driving force.

An insect, therefore, is not afraid of gravity; it can fall without danger, and can cling to the ceiling with remarkably little trouble. It can go in for elegant and fantastic forms of support like that of the daddy-long-legs.   But there is a force which is as formidable to an insect as gravitation to a mammal.   This is surface tension.   A man coming out of a bath carries with him a film of water of about one-fiftieth of an inch in thickness.   This weighs roughly a pound.   A wet mouse has to carry about its own weight of water.   A wet fly has to lift many times its own weight and, as every one knows, a fly once wetted by water or any other liquid is in a very serious position indeed.   An insect going for a drink is in as great danger as a man leaning out over a precipice in search of food.   If it once falls into the grip of the surface tension of the water—that is to say, gets wet—it is likely to remain so until it drowns. A few insects, such as water-beetles, contrive to be unwettable, the majority keep well away from their drink by means of a long proboscis.

Of course tall land animals have other difficulties. They have to pump their blood to greater heights than a man and, therefore, require a larger blood pressure and tougher blood-vessels.   A great many men die from burst arteries, especially in the brain, and this danger is presumably still greater for an elephant or a giraffe.   But animals of all kinds find difficulties in size for the following reason.   A typical small animal, say a microscopic worm or rotifer, has a smooth skin through which all the oxygen it requires can soak in, a straight gut with sufficient surface to absorb its food, and a simple kidney.   Increase its dimensions tenfold in every direction, and its weight is increased a thousand times, so that if it is to use its muscles as efficiently as its miniature counterpart, it will need a thousand times as much food and oxygen per day and will excrete a thousand times as much of waste products.

Now if its shape is unaltered its surface will be increased only a hundredfold, and ten times as much oxygen must enter per minute through each square millimetre of skin, ten times as much food through each square millimetre of intestine. When a limit is reached to their absorptive powers their surface has to be increased by some special device. For example, a part of the skin may be drawn out into tufts to make gills or pushed in to make lungs, thus increasing the oxygen-absorbing surface in proportion to the animal's bulk. A man, for example, has a hundred square yards of lung. Similarly, the gut, instead of being smooth and straight, becomes coiled and develops a velvety surface, and other organs increase in complication. The higher animals are not larger than the lower because they are more complicated. They are more complicated because they are larger. Just the same is true of plants. The simplest plants, such as the green algae growing in stagnant water or on the bark of trees, are mere round cells. The higher plants increase their surface by putting out leaves and roots. Comparative anatomy is largely the story of the struggle to increase surface in proportion to volume.

Some of the methods of increasing the surface are useful up to a point, but not capable of a very wide adaptation. For example, while vertebrates carry the oxygen from the gills or lungs all over the body in the blood, insects take air directly to every part of their body by tiny blind tubes called tracheae which open to the surface at many different points. Now, although by their breathing movements they can renew the air in the outer part of the tracheal system, the oxygen has to penetrate the finer branches by means of diffusion. Gases can diffuse easily through very small distances, not many times larger than the average length travelled by a gas molecule between collisions with other molecules. But when such vast journeys—from the point of view of a molecule—as a quarter of an inch have to be made, the process becomes slow. So the portions of an insect's body more than a quarter of an inch from the air would always be short of oxygen. In consequence hardly any insects are much more than half an inch thick. Land crabs are built on the same general plan as insects, but are much clumsier. Yet like ourselves they carry oxygen around in their blood, and are therefore able to grow far larger than any insects. If the insects had hit on a plan for driving air through their tissues instead of letting it soak in, they might well have become as large as lobsters, though other considerations would have prevented them from becoming as large as man.

Exactly the same difficulties attach to flying. It is an elementary principle of aeronautics that the minimum speed needed to keep an aeroplane of a given shape in the air varies as the square root of its length. If its linear dimensions are increased four times, it must fly twice as fast. Now the power needed for the minimum speed increases more rapidly than the weight of the machine. So the larger aeroplane, which weighs sixty-four times as much as the smaller, needs one hundred and twenty-eight times its horse-power to keep up. Applying the same principles to the birds, we find that the limit to their size is soon reached. An angel whose muscles developed no more power weight for weight than those of an eagle or a pigeon would require a breast projecting for about four feet to house the muscles engaged in working its wings, while to economize in weight, its legs would have to be reduced to mere stilts. Actually a large bird such as an eagle or kite does not keep in the air mainly by moving its wings. It is generally to be seen soaring, that is to say balanced on a rising column of air. And even soaring becomes more and more difficult with increasing size. Were this not the case eagles might be as large as tigers and as formidable to man as hostile aeroplanes.

But it is time that we passed to some of the advantages of size. One of the most obvious is that it enables one to keep warm. All warm-blooded animals at rest lose the same amount of heat from a unit area of skin, for which purpose they need a food-supply proportional to their surface and not to their weight. Five thousand mice weigh as much as a man. Their combined surface and food or oxygen consumption are about seventeen times a man's. In fact a mouse eats about one quarter its own weight of food every day, which is mainly used in keeping it warm. For the same reason small animals cannot live in cold countries. In the arctic regions there are no reptiles or amphibians, and no small mammals. The smallest mammal in Spitzbergen is the fox. The small birds fly away in the winter, while the insects die, though their eggs can survive six months or more of frost. The most successful mammals are bears, seals, and walruses.

Similarly, the eye is a rather inefficient organ until it reaches a large size. The back of the human eye on which an image of the outside world is thrown, and which corresponds to the film of a camera, is composed of a mosaic of 'rods and cones' whose diameter is little more than a length of an average light wave. Each eye has about half a million, and for two objects to be distinguishable their images must fall on separate rods or cones. It is obvious that with fewer but larger rods and cones we should see less distinctly. If they were twice as broad two points would have to be twice as far apart before we could distinguish them at a given distance. But if their size were diminished and their number increased we should see no better. For it is impossible to form a definite image smaller than a wave-length of light. Hence a mouse's eye is not a small-scale model of a human eye. Its rods and cones are not much smaller than ours, and therefore there are far fewer of them. A mouse could not distinguish one human face from another six feet away. In order that they should be of any use at all the eyes of small animals have to be much larger in proportion to their bodies than our own. Large animals on the other hand only require relatively small eyes, and those of the whale and elephant are little larger than our own.

For rather more recondite reasons the same general principle holds true of the brain. If we compare the brain-weights of a set of very similar animals such as the cat, cheetah, leopard, and tiger, we find that as we quadruple the body-weight the brain-weight is only doubled. The larger animal with proportionately larger bones can economize on brain, eyes, and certain other organs.

Such are a very few of the considerations which show that for every type of animal there is an optimum size. Yet although Galileo demonstrated the contrary more than three hundred years ago, people still believe that if a flea were as large as a man it could jump a thousand feet into the air. As a matter of fact the height to which an animal can jump is more nearly independent of its size than proportional to it. A flea can jump about two feet, a man about five. To jump a given height, if we neglect the resistance of the air, requires an expenditure of energy proportional to the jumper's weight. But if the jumping muscles form a constant fraction of the animal's body, the energy developed per ounce of muscle is independent of the size, provided it can be developed quickly enough in the small animal. As a matter of fact an insect's muscles, although they can contract more quickly than our own, appear to be less efficient ; as otherwise a flea or grasshopper could rise six feet into the air.

And just as there is a best size for every animal, so the same is true for every human institution. In the Greek type of democracy all the citizens could listen to a series of orators and vote directly on questions of legislation. Hence their philosophers held that a small city was the largest possible democratic state. The English invention of representative government made a democratic nation possible, and the possibility was first realized in the United States, and later elsewhere. With the development of broadcasting it has once more become possible for every citizen to listen to the political views of representative orators, and the future may perhaps see the return of the national state to the Greek form of democracy. Even the referendum has been made possible only by the institution of daily newspapers.

To the biologist the problem of socialism appears largely as a problem of size. The extreme socialists desire to run every nation as a single business concern. I do not suppose that Henry Ford would find much difficulty in running Andorra or Luxembourg on a socialistic basis. He has already more men on his pay-roll than their population. It is conceivable that a syndicate of Fords, if we could find them, would make Belgium Ltd. or Denmark Inc. pay their way. But while nationalization of certain industries is an obvious possibility in the largest of states, I find it no easier to picture a completely socialized British Empire or United States than an elephant turning somersaults or a hippopotamus jumping a hedge.

DARWINISM TO-DAY

'Darwinism is dead'
—Mr. H. BELLOC.

THERE is a singularly universal agreement among biologists that evolution has occurred; that is to say, that the organisms now living are descended from ancestors from whom they differ very considerably. One or two, including a distinguished Jesuit entomologist, try to narrow down its scope, but so far as I know none deny it. To do so it would be necessary either to affirm that fossils were never alive, but created as such, presumably by the devil as stumbling blocks ; or that species were wiped out, and their successors created, on a slightly fantastic scale. For example, the members of one single genus of sea urchins would have to have been destroyed and replaced by barely distinguishable successors some dozens of times during the course of the deposition of the English chalk. This is a *reductio ad absurdum* of a view which was tenable when only a few groups of extinct organisms belonging to very different epochs were known. But if evolution is admitted as a historical fact it can still be explained in many different ways.

The iguanodon has been replaced by the sheep and cow, the Austrian empire by the succession states. Some few people will attribute both these events to the direct intervention of the Almighty, a few others to the mere interaction of atoms according to the laws of physics and chemistry. Most will adopt some inter-

mediate point of view. We have therefore to ask ourselves whether evolution shows signs of intelligent or
even instinctive guidance ; and if not, whether it can be
explained as the outcome of causes which we can see at
work around us, and whose action is fairly intelligible.

Popular ideas of evolution are greatly biased by the
fact that so much stress is laid on the ancestry of such
animals as men, horses, and birds, which are, according
to human standards of value, superior to their ancestors.
We are therefore inclined to regard progress as the rule
in evolution. Actually it is the exception, and for
every case of it there are ten of degeneration. It is
impossible to define this latter word accurately, but I
shall use it to cover cases where an organ or function
has been lost without any obvious corresponding gain,
and in particular the assumption of a parasitic or sessile
mode of life.

To take an obvious example, the birds were almost
certainly derived from a single ancestral species which
achieved flight. This achievement was followed by a
huge outbreak of variation which has given us the
thousands of bird species alive to-day. The essential
step was made once, and once only. But the power
of flight has been lost on many different occasions, for
example by the ostrich and its allies, the kiwi, the dodo,
the great auk, the penguin, the weka, the pakapo (a
flightless parrot), and so on. Only the auk and penguin
converted their wings into flippers and may, perhaps,
be absolved from the stigma of degeneracy. Similarly,
hundreds of groups have independently taken to
parasitism, and in many cases very successfully. On
the average, every vertebrate harbours some dozens of

parasitic worms, whose remote ancestors were free-living. Blake asked somewhat doubtfully of the tiger,

'Did he who made the lamb make thee?'

The same question applies with equal force to the tapeworm, and an affirmative answer would clearly postulate a creator whose sense of values would not commend him to the admiration of humanity.

But in spite of this he might be an intelligent being. Now it is perhaps the most striking characteristic of an intelligent being that he learns from his mistakes. On the hypothesis of an intelligent guidance of evolution we should, therefore, expect that when a certain type of animal had proved itself a failure by becoming extinct the experiment of making it would not be tried repeatedly. Yet this has often happened. Both reptiles and mammals have on numerous occasions given rise to giant clumsy types with from one to six short horns on the head. One remembers Triceratops, Dinoceras, Titanotherium, and others. Not only did they all become extinct, but they did not even, like some other extinct animal types, flourish over very long periods. And the rhinoceros, which represents the same scheme among living animals, was rapidly becoming extinct even before the invention of the rifle. But all these animals were evolved independently. Among the titanotheres alone, eleven distinct lines increased in size, developed horns, and perished.

Two or three such attempts would have convinced an intelligent demiurge of the futility of the process. That particular type of mistake is almost the rule in vertebrate evolution. Again and again during Mesozoic times, great groups of reptiles blossomed out into an inordinate increase of bulk, a wild exuberance of scale and spine, which invariably ended in their extinction. They doubtless enjoyed the satisfaction of squashing a number of our own ancestors and those of the existing reptilian groups, who seem to have been relatively small and meek creatures.

It would appear, then, that there is no need to postulate a directive agency at all resembling our own minds, behind evolution. The question now remains whether it can be explained by the so far known laws of nature. In the discussion which follows we do not, of course, raise the questions as to how life originated, if it ever did ; or how far the existence of an intelligible world implies the presence behind it of a mind.

Darwin recognized two causes for evolution, namely, the transmission to the descendants of characters acquired by their ancestors during the course of their lives, and selection. He laid more stress on the latter and was the first to point out its great importance as a cause of evolution ; but—as might be noted by certain anti-Darwinian writers, were they to read Chapter I. of the *Origin of Species*—he was far from neglecting the former. Nevertheless, thanks in the main to Weissmann, the majority of biologists to-day doubt whether acquired characters are transmitted to the offspring. A vast amount of work has been done to demonstrate the possible effect on an organ of its use or disuse throughout many generations. To take a recent example, Payne bred *Drosophila*—a fly which tends to move towards light—in darkness for seventy-five generations. At the end of that time no visible change had occurred in the eyes ; and when one thousand such flies were given the opportunity of moving toward a light, no change was found from the normal, either in the proportion which moved within a minute, or in the average rate at which they moved. The majority of the experiments on the inheritance of the effects of use and disuse lead to equally negative results.

Some of the apparently successful experiments can be explained by selection. For example, wheat taken from Scandinavia to Central Europe and brought back again after some years was found to germinate earlier than its ancestors, and the results were attributed to the effects of earlier germination in a warmer climate. But whereas in Scandinavia the earliest germinating shoots would tend to be nipped by frost, in a warmer climate they would get a start over the later and be represented in greater numbers in each successive generation. Hence, if there was any inheritable variation in time of sprouting, selection would occur, and the wheat as a whole would sprout earlier.

Nevertheless a certain number of cases remain which can hardly be explained away in this manner, nor by the transmission of micro-organisms. It must be remembered that, however many experiments fail, it is always possible that the effects of use and disuse may be impressed on a species at a rate not susceptible of experimental verification, yet rapid enough to be of importance in geological time. But the acceptance of this principle, and in particular of the corollary that instinct is in part inherited memory, raises difficulties at least as great as it solves. The most perfect and complex instincts are those of the workers of social insect species, such as bees and termites. Now a worker bee is descended almost if not quite exclusively from queens and drones. None, or extremely few, of her ancestors have been workers. If therefore memory were inherited, the instincts of workers should slowly alter in such a way that their behaviour came to resemble that of sexual forms, and insect societies should be inherently unstable—whereas in fact they appear to date back for at least twenty million years.

The case for natural selection is far stronger. Let us first be clear what is meant by this phrase. Among the offspring of the same parents variations occur. Some of these are due to accident or disease and are not transmitted to the next generation, others are inheritable. For example, a single litter of rabbits often contains both coloured and white members. If the whites are bred together, they produce only white young. The coloured will produce a majority like themselves and a proportion of whites. That is to say, both characters are more or less markedly inherited. If now the animals bearing one inheritable character produce on the whole more offspring which survive to maturity in the next generation, the proportion of the population bearing that character will tend to increase. The phrase ' survival of the fittest ' is often rather misleading. It is types and not individuals that survive.

Of two female deer, the one which habitually abandons its young on the approach of a beast of prey is likely to outlive one which defends them ; but as the latter will leave more offspring, her type survives

even if she loses her life. Hence, in so far as courage
and maternal instinct are inherited they will tend to
survive, even if they often lead to the death of the
individual. Of course, the fact that nature favours
altruistic conduct in certain cases does not mean that
biological and moral values are in general the same.
As Huxley pointed out long ago, this is by no means
the case, and an attempt to equate moral and biological
values is a somewhat crude form of nature worship.
But that is not to say that the moralist can neglect
biological facts.

The assertion is still sometimes made that no one has
ever seen natural selection at work. It is therefore
perhaps worth giving in some detail a case recently
described by Harrison. About 1800 a large wood in
the Cleveland district of Yorkshire containing pine
and birch was divided into two by a stretch of heath.
In 1885 the pines in one division were replaced by
birches, while in the other the birches were almost
entirely ousted by pines. In consequence the moth
Oporabia autumnata, which inhabits both woods, has
been placed in two different environments. In both
woods a light and a dark variety occur, but in the pine
wood over ninety-six per cent. are now dark, in the birch
wood only fifteen per cent. This is not due to the
direct effect of the environment, for the dark pine wood
race became no lighter after feeding the caterpillars
on birch trees in captivity for three generations, nor can
the light form be darkened by placing this variety on
pines. The reason for the difference was discovered
on collecting the wings of moths found lying about in
the pine wood, whose owners had been eaten by owls,

bats, and night-jars. Although there were more than twenty-five living dark moths to each light one, a majority of the wings found were light coloured. The whiter moths, which show up against the dark pines, are being exterminated, and in a few more years natural selection will have done its work and the pine wood will be inhabited entirely by dark coloured insects. Naturalists are at last beginning to realize the importance of observations of this kind, but they require a combination of field observations with experiment such as is too rarely made.

Now it is clear that natural selection can only act when it finds variations to act on. It cannot create them, and critics have therefore objected that it cannot really be said to create a new species. It would follow from this line of reasoning that a sculptor who hews a statue from a block of marble has not really made the statue. He has merely knocked away some chips of stone which happened to be round it! Natural selection is creative in the same sense as sculpture. It needs living organisms exhibiting inheritable variations as its raw material. It is not responsible for the existence of organisms, but it remains to be shown that without it organisms would display any tendency to evolve.

Of course, if variation is biased in some one direction, a new problem arises. Variation has been adequately studied only during the last twenty years, and it is necessary to digress on the results of this study. Most inheritable variations which have been investigated are transmitted according to Mendel's laws, except that complete dominance is rather rare. That is to say, they are due to the handing on from parent to offspring of a unit which we call a gene, and which is a material structure, located at a definite point in the nucleus of the cell and dividing at each nuclear division. Characters which appear to vary continuously generally prove on analysis to be due to the interaction of a number of such genes. Now apart from non-inheritable ' fluctuations ' due to the environment, there are two distinct types of variation. The first and commonest kind is caused by a mere reshuffling of genes. If we mate a black and white rabbit with a blue angora (long-haired) doe, the offspring, if the parents were pure bred, will be black short-haired rabbits ; but among their children, if they are mated together, will appear an outburst of variation. Black, blue, black and white, blue and white rabbits will appear, some of each kind having short hair, some long, due to a reshuffling of the genes contributed by the parents. This sort of variation obeys the laws of chance, and selection will only be able to pick out one most favoured combination, say short-haired blue rabbits. Almost all variation in the human race is due to this cause.

But there is another and far rarer kind of variation, known as mutation, which consists in the origin of a new gene. I might breed a million rabbits without getting more than a dozen or so well-marked mutations. But the sort of mutations I should expect would be on more or less familiar lines. I should not be surprised if I got an outbreak of hereditary baldness, or came on a new race of rabbit with pink eyes and a yellow coat, for these types have arisen in mice ; but I should be dumbfounded if one of my rabbits developed hereditary horns, and still more so if feathers were to appear !

As a matter of fact, there is a marked parallelism between the new genes which have arisen in nearly related species ; and this is intelligible because the structure of their nuclei is similar, and the changes likely to occur in them are therefore also similar. New genes appear to arise as the result of accidents—that is to say, causes which are no doubt determined by the laws of physics, but are no more the concern of the biologist than those governing the fall of a chimney-pot, which has been known to alter the shape of a human head, though not in an inheritable manner. Mutations have been provoked in mice and flies by mild injury of the germ plasm with X-rays. The vast majority of mutations are harmful, resulting in an impairment of some structure or function, and are eliminated by natural selection. Others are neutral. In a fly of which some tens of millions have been bred in laboratories, over four hundred mutations have occurred, some of them on many different occasions. Only two have yielded types as healthy as the normal. Advantageous mutations are still rarer—that is why evolution is so slow. But they do occur.

On a Sumatran tobacco plantation a new type of tobacco plant, due to a mutation inherited on Mendelian lines, arose suddenly. It was found that the new variety, though no better off than its ancestors in Sumatra, gave distinctly better crops in a cool climate. If it had arisen in the wild state it would have enabled the tobacco plant to extend its northerly range and form a new subspecies. It must be remembered that a mutation which in most circumstances would be disadvantageous, may be useful in a special environment. Wingless varieties of normally winged insects are common on small oceanic islands, though by no means universal. Mutations causing loss of wings are also common in the laboratory. It is clear that after an island has been colonized by a winged insect carried by the wind from an adjoining continent, hereditary loss of wings, if not accompanied by degeneration of other structures, will be of value in preventing its successors from being blown out to sea.

It is clear, then, that in mutations of this type we have a means by which subspecies may be formed in nature, and there is strong evidence that they have been so formed. For example, the three varieties of the black rat, which have different geographical distributions, differ from one another by single genes quite similar to those which arise by mutation in the laboratory. But there is no evidence at all that mutations are biased in a direction advantageous to the species. The possibilities of mutation do, however, limit the directions in which a species can evolve. Whether it will do so along any of the lines thus laid open to it depends on natural selection. In some cases, as among flowering plants, a good many species seem to be neither better nor worse off than their ancestors—and therefore to owe their origin primarily to variation. However, a slight change in leaf or flower form can hardly be called evolution.

In many cases a change in one character will only be of advantage to a species if some other varies simultaneously in the same direction. This has been used as an argument against natural selection. But in the first place, although one gene may affect one structure

only or mainly, others will modify a whole group.
Thus of the genes which alter the wing of the fly
Drosophila some have little effect elsewhere, some also
affect the balancers (rudiments of the second wing
pair), others the legs, and so on. A mutation will,
therefore, often be found to kill the two birds with one
stone, so to speak. Should this be impossible, selection
can still work.

Suppose it is to the advantage of an animal that two
structures A and B—say bones—should increase together, but that variations in them are inherited independently. We can classify the animals according as
the two are of less than the average size, greater than
the average or about equal to it. So that we get nine
classes in all. Those in which the two are unequally
developed will be at a disadvantage, only where both
are increased will there be any gain. Putting the
number of the normal type surviving at 100, we should
get survival rates somewhat as follows :—

	A+	A=	A−
B+	101	98	96
B=	98	100	98
B−	96	98	99

where the figure 101 represents the fact that animals
with both A and B increased have a one per cent.
better chance of survival than the normal. It will be
seen that the A— and B— groups will tend to die out,
so that both structures will increase in size.

To my mind the most serious argument against
selection on these lines is that it does not explain the
origin of interspecific sterility, except where it is due
to external causes such as differences of size or breeding

time. It is on these grounds that Bateson, a thorough believer in evolution, has criticized natural selection. But I have pointed out elsewhere,[1] a difference of a single gene between two animals may cause the production of an excess of one sex on crossing, as occurs in fowl-pheasant and cow-bison crosses ; and several such genes may well cause complete sterility.

Moreover, there is a second type of inheritable variation, leading to a change in the chromosome number, which causes inter-varietal sterility, often without a very marked change in external characteristics. This is quite common in plants, less so in animals. Although, therefore, the problem of interspecific sterility is serious, we are already well on the way to solving it.

We must now turn to the palaeontological evidence. In a few groups we can trace the course of evolution in some detail. Thus we know over five thousand species of ammonites, and nearly two hundred of extinct horses. In the horses, advance took place along several parallel lines, only one of which has left living descendants. In each line the toes were gradually reduced from three to one, while the molar teeth increased in length and complexity. When in the past we find two different species competing in the same area, one is usually further on the road towards a single toe, the other towards a long molar. We know that these two characters were of value, because we find fossils in which the thin lateral toes—reduced to mere vestiges in the modern horse—had been broken during the animal's life, as shown by subsequent healing. We also find that in the more primitive types the teeth were

[1] *Journal of Genetics*, vol. 12.

often worn down to the roots, leading to death from starvation. Hence for two species to compete equally their advantages in these two respects must be balanced, since species combining both advantages—as does the modern horse—would oust those possessing one only. Evolution in the cases where the evidence is most complete is known to have been very gradual. Such large changes as those produced by most genes so far studied were rare in evolution. This is natural enough. Geneticists have concentrated their attention on genes which produce striking effects. Now, however, that they are beginning to look for those causing very small effects only, and often apparently continuous variation, they are finding them.

A more serious objection is that rudimentary characters sometimes appear which can be of no use to their owners, but only become so on further development some thousands of years later. This is almost certainly true and is at first sight fatal to the selection hypothesis. But it can be met along several lines. A change in one organ, as Darwin pointed out, generally carries with it a change in others. Hence an increase in the complexity of one molar brought about by natural selection may cause the beginning of a new cusp in its neighbour. This cusp will at first be useless, but as it increases selection will begin to act on it also, so that the process will gather momentum until we arrive at the extremely complex grinders of the elephant or horse. Moreover, we can trace just the same gradual beginnings of apparently quite useless organs, the excessive skeletal outgrowths which have been the harbingers of extinction in many animal groups, both vertebrate and invertebrate. If we knew more about these creatures' soft parts we could perhaps elucidate these problems. Some light is thrown on them by recent work of J. S. Huxley and others. They have shown that, in certain animals, growth of the whole body leads to disproportionate growth of one part. Thus in a group of crabs, whenever the body doubles in weight, the large claw increases three times, until it finally becomes almost as large as the rest of the animal. Any cause promoting growth of the whole body, therefore, leads to a disproportionate growth of the claw. And such a cause is to be found in competition within the species, more especially the competition between males for females by fighting, as is common among mammals, rather than display, as seems to be the custom with many birds.

Still the possibility of some deeper underlying cause of evolution is often suggested by the study of a whole great group, such as the ammonites, which furnish the best available material, for the following reasons : They were sea-beasts, hence their shells were preserved far better than the skeletons of land animals. The number of known fossil species is nearly double that of living mammals. Their shells tell us of their development, for the whorls formed by the young animal are preserved in the middle of the complete structure. Finally, their history is over. The last of them died in Eocene times, twenty million or more years ago.

The earliest forms were often not coiled at all and always had very simple patterns on the sutures between different shell chambers ; and their descendants still made these simple patterns in the embryonic stages.

In the great ages of ammonites during the first two-thirds of the Mesozoic era, the most complex ornamentation was generally made by the adult animal. But as time went on, it showed a tendency to slur its work. The most complex patterns were made by the half-grown creatures, and in Cretaceous times the adult shells were sometimes even uncoiled, as in the very earliest forms. Now this ' second childhood ' occurred independently in some scores of different lines of descent, always as a prelude to extinction. In other groups the same phenomenon may be observed, though the stigmata of degeneration are different.

This degenerative process is often described as the old age of a race, but we must remember that this phrase is only a metaphor. Some very obvious explanations for it are as follows :—

A step in evolution in any animal group is followed by an evolutionary advance on the part of their parasites. When our fish ancestors came out of the water, they lost their louse-like crustacean parasites ; and it was only after some time that insects can have taken their places ; and later still that micro-organisms such as those of malaria and typhus were evolved, which pass part of their life-cycle in insects and part in vertebrates. So the apparent degeneration of a group may only mean that evolution of their enemies has caught up with their own. Again, specialization—while it leads to temporary prosperity—exposes a species to extinction or at least to very unfavourable conditions when its environment alters. A small change of climate will lead to a disappearance of forests over a wide area, and with them of most of the animals highly adapted to life in them, such as squirrels, woodpeckers, wood-eating beetles, and so forth. A few, like our own ancestors, adapted themselves to a new environment; but the majority, and all the more highly specialized, died out, the new population of the area being recruited from among the less well adapted forms. Also, as pointed out above, competition within the species, man included, may lead to results desirable for a few individuals, but most undesirable for the species as a whole.

To my mind the closest analogy to the evolution of a given group is the history of the art and literature of a civilization. The clumsy primitive forms are replaced by a great variety of types. Different schools arise and decline more or less rapidly. Finally a period of decline sets in, characterized by archaism like that of the last ammonites. And it is difficult not to compare some of the fantastic animals of the declining periods of a race with the work of Miss Sitwell, or the clumsy but impressive with that of Epstein. The history of an animal group shows no more evidence of planning than does that of a national literature. But both show orderly sequences which are already pretty capable of explanation.

To sum up, no satisfactory cause of evolution other than the action of natural selection on fortuitous variations has ever been put forward. It is by no means clear that natural selection will explain all the facts. But the other suggested causes are unverified hypotheses, while selection can be observed by those who take sufficient trouble. Some of the alleged causes, moreover, are difficult to reconcile with the facts of palaeontology and genetics. The evidence as to the earth's age from radioactive minerals shows that about six hundred million years have elapsed since the first known fossils were laid down, and perhaps twice as long since life appeared on the earth. This is a larger time than the early supporters of Darwin demanded, and seems long enough to satisfy any quantitative objections as to the slowness of evolution. There are qualitative objections, such as those connected with the origin of consciousness. But consciousness arises anew in every human being. Its first origin on the earth presents no more and no less mystery than its last.

Finally, no facts definitely irreconcilable with Darwinism have been discovered in the sixty years and more that have elapsed since the formulation of Darwin's views. Such a fact would be, for example, a convergence in the course of geological time of members of two or more groups to form a single species. Actually, we observe the convergence of forms as we go down and not up a geological series. And there have been quite enough anti-Darwinian palaeontologists to have seized on such a case had it existed.

As an explanation of evolution Darwin's ideas still hold the field to-day, and subsequent work has necessitated less modification of them than of those of his contemporaries in physics and chemistry. Just as physiology has found no case of interference with the order of nature as revealed by physics and chemistry, the study of evolution has brought to light no principle which cannot be observed in the experience of ordinary life and successfully submitted to the analysis of reason.

ENZYMES

THE chemical processes which take place in a living animal or plant are just as characteristic as its form or behaviour. Yet, taken one at a time, they can often be imitated by artificial means. Rhumbler made an artificial 'cell' which would absorb a glass thread covered with sealing-wax, remove the wax and spit out the thread. Hammond made a motor 'dog' with selenium 'retinae' which would follow a light or a white object. And similarly we can imitate many of the chemical reactions which take place in the cell, though often we require rather violent means, such as heat or the application of strong acids. What is characteristic of life is not the individual details of structure or behaviour, but the way in which they cohere to form a self-regulating and self-preserving whole.

When we succeed in investigating the details of a chemical process in the cell we generally, if not always, find that it is determined by the presence of an enzyme or ferment, which can be more or less completely separated from the rest of the cell and is not alive. In favourable cases we can break up the cell, and, by a series of processes not utterly unlike those employed in extracting the rarer metals from their ores, except that no heat is used, obtain one of the enzymes in a fairly pure state. A solution of cane sugar in water is stable, but if we warm it with a strong acid it breaks up into the mixture of sugars found in honey. This can also be done by enzymes found in plant cells, the saliva of bees, and our own intestines. The most active enzyme preparations, when dissolved in water, will break up ten times their weight of sugar per second and, as far as we know, will continue to do so indefinitely. Certainly they can break up more than a million times their own weight without wearing out.

The early workers on enzymes believed that a little ' vital force ' resided in these particles, a belief analogous to that of primitive men in the magical nature of their own tools. But enzymes are certainly not alive. They do not reproduce themselves nor adapt themselves to changes in their environment. They are simply the tools of the cell. Their action is similar to that of inorganic catalysts, such as finely divided nickel or platinum, which are used in industry to speed up many chemical reactions, or to allow them to occur at a lower temperature than would be possible in their absence. And although we have not yet been able to make an enzyme artificially, and shall not be able to do so for many years, we are gradually elucidating their chemical composition. Many of them would seem to be proteins, and it has been suggested that the protein of certain cells consists almost wholly of enzymes.

One of the most characteristic things about an enzyme is its specificity. The enzyme which digests cane sugar will not touch milk sugar or malt sugar, and conversely. Enzymes have been compared to keys which will only open certain locks. One might go further and say that they are Yale keys. Many molecules which are attacked by them are asymmetrical, as is shown by the asymmetry of their crystals, and the fact that their solutions rotate the plane of polarized light passed through them. We can often make the mirror images of these molecules, and we then find that the corresponding enzymes will only attack them slowly if at all. On going through the looking-glass, Alice would have found her digestive enzymes of no more use on the looking-glass sugars than her Yale key on the looking-glass locks. This asymmetrical behaviour is not, however, peculiar to enzymes. It has been found to hold good for the action of other catalysts, including some synthetic substances.

Our knowledge of enzymes has so far had rather little direct application. Those first studied were the ones found in digestive juices, which break up our food into readily absorbable substances, and are unique in that their action normally takes place outside the cell where they are formed. Preparations of these can readily be made, and one of them, rennet, has long been employed in dairies to clot milk. But no very great success has attended the treatment of various forms of digestive trouble by administering enzymes. The main reason for this is that indigestion is seldom due to a shortage of them. So at the present day, apart from rennet and an extract of the pancreas used in the partial digestion of hides to furnish certain kinds of leather, preparations of the digestive enzymes are of more benefit to their sellers than their buyers. For example, a mixture containing pepsin is widely advertised as a tooth-paste, although pepsin is inactive in such fluids as saliva, a fairly strong acid being needed to make proteins susceptible to its action.

Several diseases and abnormalities are due to the absence of an enzyme, but we cannot cure albinism, for example, by injecting tyrosinase as we cure diabetes with insulin. For insulin appears to have a small enough molecule to allow it to get through the walls of the cells, whilst enzyme molecules are too large for this to be at all easy, and they seem generally to be manufactured on the spot where they will be used. So far, therefore, our efforts in medicine are directed rather to enabling such enzymes as exist in the cell to act more efficiently than to supplying them in their absence. We are now, however, beginning to study the enzymes not only of bacteria but of cancer cells, which seem to be slightly but significantly different from those of normal tissues.

In industry we generally find it better to use enzymes in the cell than out of it. Yeast makes alcohol from sugar much more rapidly than does any extract containing the three or four enzymes concerned in the process. But it does a good deal more. The yeast cell breaks up proteins as well as sugar, and from these it forms the fusel oil which not only gives an alcoholic liquor much of its characteristic taste and smell, but also in many cases makes it a great deal more poisonous than its alcohol content would lead one to suppose. It is quite possible, therefore, that if the liquor trade has a future it may be on the lines of utilizing enzyme preparations rather than living yeast cells. When yeast cells are given cane sugar, they break it up into simpler sugars before making these latter into alcohol.

The enzyme which they employ for this purpose may easily be separated from those concerned in fermentation. Now the new sugars formed are both more tenacious when wet, and more retentive of water, than cane sugar, and therefore better material for making sweets. So the use of yeast invertase is spreading in the American candy trade, and if prohibition has raised the consumption of candy, it has not been wholly disadvantageous to the yeast plant.

Perhaps the biggest field for the commercial application of enzymes lies elsewhere. Plants make sugar from carbon dioxide, water, and sunlight. A few, like the potato, store most of this in starch, or some other form that we can digest. But the majority of them convert it into cellulose, the main constituent of wood. They commonly make a little sugar to attract bees or birds, but man is the first mammal who has seriously befriended plants, and agriculture is so recent that only a few of the more enterprising among them have had time to vary so as to bribe him with food. The cow and horse can no more digest cellulose than ourselves. No animal nearer to us than a snail can make the enzymes requisite for even a partial digestion of it. But an ungulate is a co-operative society. It consists of the mammal which forms the façade, and some millions of millions of bacteria engaged in breaking down cellulose. The products which they form from it are largely digestible by the horse or cow, but would be unpalatable, if not harmful, to man. However, one of the intermediate stages in their production from cellulose is an easily digestible sugar. When—not if— we can separate the cellulose-splitting enzymes from those which break up the sugar further, we shall be in a position to convert wood pulp or hay quantitatively into human food.

This is one of the facts which render dubious all prophecies as to over-population. The upper limit to human numbers is not set by any facts of nature, but by human ignorance and inadaptability.

VITAMINS

A VERY large amount of nonsense is written on the subject of vitamins, and some good purpose may be served by attempting to summarize what is known at the moment of writing, and may be out of date when this article is read. For a long time in the past it had been understood that scurvy was due to a special type of monotonous diet, and could be cured by small amounts of certain fruits and vegetables, or larger quantities of fresh meat. Later the same was proved for beri-beri and suspected for rickets and other diseases. Meanwhile the problem of the ideal diet had been largely solved in the nineteenth century. It was shown that one of the first necessities was a sufficient fuel value in the food. When burnt it must be able to provide the energy needed by the body. But only a few substances, namely, carbohydrates, fats, and proteins, will serve as energy sources, and not too much must be fat. A certain minimum of proteins, which must be of the right quality, is needed for repairs of the tissues. Elements like iron and calcium must also be present in small amounts.

In the early twentieth century the problem was attacked from two points of view. First of all, a number of workers dealt with the question of what had to be added to an otherwise complete diet in order to prevent a given deficiency disease. This may be called the analytical method. The converse or synthetic method, of which Hopkins was the most successful exponent, asked the more ambitious question, ' Can we make a complete diet out of substances of known chemical composition, and, if not, what must we add to it to make it satisfactory ? ' These lines of research have now converged. Besides the discovery of the vitamins they have led, among other things, to a knowledge of what chemical properties in a protein are needed to make it a satisfactory constituent of a diet, and to the proof that most cases of goitre are due to iodine deficiency, while traces of zinc are almost certainly necessary in the diet.

It now seems that besides proteins, carbohydrates, fats and inorganic substances, at least five rather complex organic bodies are needed. Hopkins called them accessory food factors. Funk, who had obtained a preparation containing a good deal of one of them and believed that he had got it pure, called it vitamine ; and this name not only caught on, but was applied to the other accessory factors.

The general method of research is as follows. Two groups of very similar animals, usually rats, are fed on diets which differ in only one particular, say, the addition to one of them of a little killed yeast ; and the difference between the two groups is observed. The group on the adequate diet may grow faster than the others, form better bones, have more young, or what not. It is rarely necessary to push the experiment so far as to lead to the death of the group on a deficient diet, and they often look quite healthy and behave normally. Indeed, since so many pet rats receive very unsatisfactory food the most striking thing to a casual observer may be the robust health of the animals on a really adequate diet. When the original experiment has been repeated and confirmed, and the exact amount of the added substance needed per rat per day has been determined, one proceeds to hunt down its essential constituent. ' Can it be replaced by its ashes ? ' we first ask, and when the answer is ' No ' we are sure that we are not concerned with an inorganic substance, such as iron or iodine. ' Is it soluble ? ' The answer so far has always been ' Yes,' if we choose the right solvent ; in certain cases, water, in others, ether or some other liquid which readily takes up oily substances.

We now try to purify it. For example, the substance in cod-liver oil which cures rickets is left behind when 99 per cent. or more of the oil is converted into soap and glycerine by heating it with alkali. We can thus obtain a preparation many hundred times richer in the antirachitic substance than was the original cod-liver oil. Sometimes the purification may show that where we thought we were dealing with a single substance we have really got two. For example, several workers obtained liquids very rich in the substance which prevents the convulsive seizures due to polyneuritis which occur in its absence. The cruder preparations of this substance always contain a substance (vitamin B) whose absence causes loss of weight. But the most active antineuritic preparations do not prevent this loss, so it is concluded that the antineuritic vitamin is not vitamin B. From such experiments we arrive at the following provisional list of vitamins. The lettering of A, B, and C is McCollum's, and is generally agreed on. That of the remainder is still under discussion.

A is an oily substance found in many natural fats and oils, cod-liver oil being particularly rich in it. Its absence leads to failure of growth in the young, and in both young and old to a tendency to inflammation of the eyes, and increased liability to various diseases. Night blindness is often an early symptom of its absence. It is slowly destroyed by cooking.

Vitamin B is found in a variety of foods, especially in certain portions of wheat and rice grains. Its absence causes failure of growth or loss of weight. Pellagra, a disease common in populations living mainly on maize, is probably due to an inadequate supply of it. It is soluble in water.

C is widely distributed, but certain fresh vegetables and fruit are particularly rich in it. It is soluble in water, and rather easily oxidized, hence the emphasis on freshness. Its absence causes scurvy, but a good many infantile troubles are also due to a shortage of it.

D is a waxy substance, long confused with A, and associated with it in nature. Both, for example, are found in cod-liver oil. We can make it in our own bodies, provided that we get enough ultra-violet ' light ' on our skins.

As, however, not only clothes, hats, and fogs, but even glass windows screen us from this component of the sun's rays, it is safer to be sure that our diet contains the vitamin. Children and young animals kept in inadequate light without it generally develop rickets, but may not do so if the amounts and proportions of calcium and phosphorus, the bone-forming elements, in the diet are kept exactly right. Vitamin D is formed by the action of ultra-violet radiation on ergosterol, a substance of known composition, and is the only vitamin yet obtained in a fairly pure state. Apparently this is the only dietary factor that is made by radiation, for no amount of sunlight will make up for the absence of any of the others.

E is also of an oily character, and is present in various foods, particularly wheat bran. In its absence rats grow up and live healthily, but they cannot reproduce. They can neither become fathers nor mothers. Such at least is the statement of its American discoverers, Evans, Bishop, Burr, and Sure, and their work has now been repeated in this country ; but at least one observer in the United States has obtained a contrary result. The existence of this vitamin cannot therefore be regarded as absolutely proved.

The letter F might be reserved for the antineuritic substance which accompanies B, and has very similar chemical properties to it. In the absence of B and F the very serious tropical disease of beri-beri develops in man, and the ' war dropsy ' of Central Europe was probably due to the same cause, but until these substances have been more satisfactorily separated it is not quite clear which of the two is the main preventative of any given complaint. There is further evidence which makes it quite likely that the above list is not exhaustive.

It is customary to conclude an article on this subject with an admonition to consume more fruit, fresh vegetables, or raw milk. The custom would be laudable if vitamin deficiency were the only fault to be found with our diets. But, as a matter of fact, about half the human race at the present moment is suffering from partial starvation, and the first requisite for them is to eat more of the cheapest food they can get, vitamins or no vitamins. And among those who read this article I suspect that for every one who is seriously the worse for vitamin deficiency there is another suffering from constipation due to a too digestible diet, which leaves no residue of husks and fibres, and three or four victims of bacteria and worms which they have absorbed with their food.

To take a simple example, I would sooner have my child run the risk of rickets or infantile scurvy from over-boiled milk than of tuberculosis from drinking it raw. I refer here to British milk—American is less tuberculous. Again, raw vegetables, though full of vitamin C, are an admirable vehicle for the typhoid bacillus, which is commoner in the United States than England. And they are so liable to contamination with the eggs of parasitic worms that the Strasbourg International Congress on cancer, impressed by the proof that worms may produce that disease, issued a perhaps unduly solemn warning against the consumption of salad. While, therefore, we cannot neglect the teachings of bio-chemistry in our choice of diet, we shall do ourselves no good if in the attempt to be up to date we neglect the lessons to be learnt from parasitology and bacteriology.

But if we are sure that our food is uncontaminated, and that there is enough of it, an adequate supply of vitamins is the next consideration. And nowhere is this consideration more urgent than in the case of infants, who can ask for more, but not for better, food.

MAN AS A SEA BEAST

KINGS and editors commonly speak in the first
person plural. If we all habitually did so, and
thought so, we should understand a good deal
more about how we work. For each of us is a community of about a hundred million million cells, whose
co-operation is our life. This co-operation is brought
about in part by the nervous system, and its beauty
and delicacy is apt to blind us to the fact that a great
many cells—in fact the majority—are not supplied
with nerve fibres. Their behaviour is determined by
two things, the mechanical and electrical forces exerted
on them by their neighbours, and the chemical composition of the fluid that surrounds them or is given to
them by their colleagues. How great is the importance of the non-nervous influences is shown by the
fact that the other parts of an embryo develop perfectly
before any nerves grow out to them from the brain and
spinal cord, and will continue to do so nearly normally
even if the nervous system does not develop at all. If
we may use the well-known comparison of the body
and the state we may say that most of our own citizens
are not state employees but act from economic and
other motives without any direct orders from the central
government. What is more, many of the cells in the
brain, the seat of government, are alert to the smallest
changes in their chemical environment; and react to
them by transmitting orders for some such activity as

an increase or decrease of the breathing, which will bring their environment back to their normal conditions.

If we observe single cells, such as the protozoa, bacteria, or the diatoms and other microscopic plants in sea water which are the ultimate source of almost all the nourishment of sea beasts, we find that they are often remarkably hard to keep alive. The tiniest changes in the fluid around them, especially in its alkalinity, will kill them or greatly alter their behaviour. Indeed they are quite as dependent on the presence of the right amount of potassium and calcium salts around them as on that of oxygen or food. As a matter of fact they spend a great deal of their energy in overcoming the defects of their surroundings. For example, water almost invariably leaks through the skins of fresh-water protozoa, and they require a special organ, the contractile vacuole, to expel it. Placed in salt water they only empty this quite rarely to get rid of waste products.

Our own cells are much more efficient than protozoa at their particular functions, but they require an extremely constant and artificial environment. It is the business of various organs, such as the lungs, liver, intestine, kidneys, and thyroid gland to keep it constant. In the same way a civilized man is generally far more efficient at his particular vocation than a savage, but only on condition that most of his needs are met by bakers, builders, tailors and so forth. Our internal environment is the blood, or rather its fluid part, the plasma in which the corpuscles are suspended. Some of the activities concerned in its regulation escape our consciousness. For example if the amount of sugar in it becomes too small, the liver makes fresh sugar from a starchlike substance called glycogen which is stored in the liver cells. If the amount of any soluble constituent in it becomes too great, the kidneys eliminate the excess, and so on. Sometimes however our consciousness and will are concerned. A shortage of water leads to thirst, a shortage of sugar which the liver cannot immediately remedy to hunger, a shortage of oxygen to panting, which may be so intense as to occupy our whole attention and will.

The blood plasma of many marine animals is almost the same as sea water, with the addition of a little sugar and other foodstuffs on the way from the gut to the cells, and waste products on the way from the cells to the excretory organs. A cockle's heart will continue to beat if placed in sea water, though quite a small change in its chemical composition, say a precipitation of the calcium (lime) salts, would render the sea water poisonous to it. We vertebrates have a blood plasma which has much the same composition as sea water diluted with three times its volume of fresh water. Such a liquid can safely be injected into the human veins in quite large quantities. The chemical agreement is far too striking to be a coincidence. Whereas all cells contain more potassium than sodium, the plasma contains 15 times as much sodium as potassium, the corresponding figure for sea water being 27. Similarly the ratio of sodium to calcium is 39 in plasma, 27 in the sea. With regard to magnesium the agreement is not so good. It is suggested that just as the plasma of modern marine invertebrates is very nearly sea water, so our own represents the sea of a remote period when our marine ancestors first began to develop gills impermeable to sea water. Modern fish, even those which live in the sea, have a plasma much like our own in its low salt content, so presumably it was their and our common ancestor that first effectively shut itself off from the sea. As the sea is always receiving salt from the rivers, and only occasionally depositing it in drying lagoons, it becomes salter from age to age, and our plasma tells us of a time when it possessed less than half its present salt content.

It is not only our tissue cells that lead this aquatic existence. Most marine animals, both vertebrate and invertebrate, shed their eggs and spermatozoa into the sea, and rely for fertilization on the numbers and swimming power of the latter. We have cut down our output of eggs to one or two a month, but we still continue, in contrast with many insects and crustaceans, to produce spermatozoa which have to swim great distances to their goal, and are therefore required in fantastically vast numbers. Their marine ancestry is shown by the fact that they can only live in a fluid containing much the same salts as the plasma. And after our development has started from the fusion of an egg and a spermatozoon we pass our first nine months as aquatic animals, suspended in and protected by a salty fluid medium. We begin life as salt-water animals.

There are two of our sense-organs which bear striking testimony to our marine ancestry. Under the skin of a fish are a number of tiny tubes occasionally opening to the exterior. There is a complicated system on the head, and one on each side of the body, often marked by a conspicuous stripe on the skin above it, as in trout.

These tubes contain bunches of microscopic hairs, richly supplied with nerve fibres, and far too delicate to be left on the outer surface of the body. The fishes' own movements through the surrounding water, and also local currents and vibrations in the water itself, are communicated to the fluid in the tubes, and bend the hairs over. Thus the fish learns of the speed and rhythm of the water movement in the tubes, as a cat might gauge the strength of a wind by the degree of bending of its whiskers.

Two parts of the tube system on each side of the head are deeply buried in the skull and highly specialized. One is adapted to respond to fine and rapid vibrations in the water, in fact to sounds. The other consists of three loops at right angles, the so-called semi-circular canals. These organs are only connected with the sea by a long narrow tube sometimes closed in the adult. But when the fish turns round, the water in one or more of the semi-circular canals is left behind, like the water in a glass which is suddenly rotated, and presses on the hair cells in the canal. Thus while the organs in the external system inform the fish of its movements relative to the water round it, those in the semi-circular canals are stimulated by its turning movements.

We land vertebrates have lost most of the fishes' canal system, but the two pairs of specialized organs in the head remain as our internal ear, open to the surrounding water in early embryonic life, but closed long before birth or hatching. The ear-drum and an elaborate system of tiny bones transmit aerial vibration to the water in one part of it. The corresponding vibrations of this water act on hair cells at the end of the auditory nerve fibres, and these in turn stimulate those parts of the brain concerned with hearing. When we turn our heads the swirling of the salt water in the semi-circular canals presses on the hair cells. An elaborate system of nerve fibres in the brain links them up to the muscles which move our eyeballs, and as we turn our heads our eyes turn in the opposite direction, so that the direction of our gaze is unaltered. This is a reflex action uncontrollable by the will ; in fact it is impossible to turn one's head suddenly while keeping the eyes fixed relatively to it.

The semi-circular canals can play us false. In a rotating bowl the water gradually comes to rest with regard to the bowl, *i.e.* takes up the bowl's rate of spin. The same happens to the fluid in our internal ears if we rotate uniformly. Hence the stimulus to the eye-muscles ceases and we can gaze steadily at any object rotating with us, for example the face of a partner in the pre-war type of waltz, while surrounding objects at rest cannot be fixed. When, however, the bowl or the man stops rotating the fluid does not, and the eyes execute involuntary movements which lead us to believe that everything is spinning round us. One can also become giddy in a vertical plane by turning round several times with the head bent forward and thus causing the fluid to swirl in a plane which becomes vertical when one lifts one's head up. The reflex now let loose involves the muscles of the limbs and trunk, and would be appropriate if one were falling over ; actually however it often makes us fall in the opposite direction.

In many ways a magnetic or gyrostatic compass would be a better balancing organ, but life has never used either the wheel or the magnet.

The evolution of the human body resembles that of the British constitution. It is full of relics of the past, as curious as the judges' wigs or the city companies, but for most of these vestiges a new function has been evolved.

FOOD CONTROL IN INSECT
SOCIETIES

MAN'S habits change more rapidly than his instincts. To-day we are born with instincts appropriate to our palaeolithic ancestors, and when we follow our instincts alone we behave in a palaeolithic manner. It is probable that primitive man, like a wild animal, 'knew' pretty well what was good for him in the way of food. Modern man does not, and when he does he cannot get it. Sedentary workers consume meals appropriate to hunters. Women of fashion attempt to supply the energy needed for dancing by the ingestion of large amounts of chocolate. Man, in fact, must use his reason to arrive at an appropriate diet. But the members of insect societies have solved a similar problem on instinctive and physiological lines. They have brought about the best possible division of a communal food supply by methods which, if strange and often disgusting to human minds, are as effective as any system of food control invented by man.

Let us see what are the prerequisites of a rational distribution. Apart from water, salts, and vitamins which are only required in tiny quantities, foodstuffs may be classified as carbohydrates, fats, and proteins. Carbohydrates include sugar, starch, and the like, fats embrace the chemically similar oils and waxes. Neither contain nitrogen or sulphur, and they are mainly useful as fuel; that is to say in order by combining with oxygen to give up energy which can be used by the animal for heating itself, or working its muscles and other organs. The proteins, on the other hand, are required to build up the living tissue during growth, and repair it after injury or the wear and tear of everyday life. If we compare the requirements of an animal and a motor vehicle, water serves the same function in both, of cooling and carrying away unwanted substances, carbohydrates and fats correspond to petrol, proteins to spare parts, and probably vitamins to lubricating oil. As a matter of fact proteins can act as a source of energy, just as spare woodwork for a train could be used as fuel, but most animals find them unsatisfactory as the sole source.

It is clear that a growing animal needs relatively more proteins than an adult. A baby lives on milk which an adult would instinctively supplement with starchy foods. But the baby requires, and finds in the milk, some fat and carbohydrate as fuel to keep itself warm and work its tiny muscles. The wasp grub is cold-blooded and sluggish. It requires very little but proteins. And the adult worker with its short life of intense exertion needs little protein but plenty of fuel. Hence, even though the food which the workers give to the grubs consists very largely of chewed-up flies, it contains more carbohydrate than necessary. When a worker comes to feed a grub by regurgitation from its crop the grub thanks it by secreting a drop of fluid containing sugar for which it has no use, but which is valuable fuel for an active insect.

The bees have taken things a stage further. Their sources of food are the nectar of flowers, a nearly pure solution of sugar in water, and the pollen, which consists largely of proteins. Even from the same flower one bee never collects both nectar and pollen. And in the hive the nectar is stored as honey, and the pollen separately as ' bee-bread.' The honey is used primarily as a source of energy and heat during the winter, the bee-bread along with some honey as food for the grubs. What is more remarkable is the fact that a grub gets a different mixture according to its future career.   Queens and workers come from fertilized eggs, drones from unfertilized, but the difference between queens and workers seems to be determined by the type of food given to the larva. So that in the hive food control is also birth control.

The most bizarre system of all is found among termites. These insects live almost entirely on wood, which most animals cannot digest. Strictly speaking the termites cannot do so either, but their intestines contain protozoa which can, just as the horse and cow digest their hay with the help of bacteria. There is evidence that these or other organisms in their bodies can even fix atmospheric nitrogen like the bacteria found in the roots of leguminous plants, thus dispensing with the need for proteins in the diet of their hosts. But this digestion is too slow a process to come to completion in the body of a single insect, so the partially digested excreta of one are eaten by another until the process is complete, and the final indigestible residue is also so incapable of putrefaction that it can be used for nest-building. This apparently repulsive process only corresponds to the passing on of half-digested food by one segment of our intestines to the next. A single termite has not a long enough intestine for the whole process. But it is only certain of the termites that can take part in wood digestion. Besides queens and males the termite nest usually contains several different castes of workers and soldiers with large jaws. These jaws are too clumsy to allow of wood-chewing, so the soldiers are fed by the workers with so-called saliva, as is the queen.

Termite societies therefore rest on a basis of physiological functions and of instincts, each one as complex and highly organized as those which form the basis of the relationship between a mammalian mother and her children. But alas, insect societies are no more perfect than human, and parasites can as easily find a place in an economic system determined by instinct, as in the products of intelligence, enlightened self-interest, or whatever else is at the basis of human economics. Whether the correct form of demand for food in an ant's or termite's nest is a gentle stroking of the donor, an offer of a drop of some sweet secretion, or what not, some unprincipled insect will generally be found to make it. Students of human society will compare these parasites with brewers, burglars, bolsheviks, bankers, bishops, or bookmakers according to their tastes. Occasionally they are of some value to the community, for instance by joining in its defence, generally they are useless, so far as we can see ; and often they devour not only food, but larval ants.

Humanity is engaged in the awkward passage from an instinctive to a rational choice of food. ' A little of w'ot yer fancy does yer good ' is no longer a sufficient guide for us, as it is for the insects, and we do not yet know quite enough to rush to the opposite extreme, though the experience of the war showed that a fairly strict rationing on scientific lines is already a possibility. But every month we are approaching nearer to a knowledge of the dietaries best suited for any individual case, a knowledge which will be as efficient as the instinct of the insect, and infinitely more elastic.

OXYGEN WANT

THE source of almost all the energy developed in the human body is the combination of food with oxygen. We can replace one kind of food by another, but oxygen cannot be replaced. The combination occurs in all the tissues, and both food and oxygen must be supplied to them by the blood. All organs are sooner or later damaged by want of oxygen, but the brain is by far the most sensitive. The first symptoms of oxygen lack are always mental, and five minutes of complete deprivation will kill the brain, whereas the heart will survive for as many hours. You, reader, will die of oxygen want. Your lungs, your heart, or that part of your brain which controls your respiratory muscles will cease to play its part in oxygen supply, and the energy transformations which make up your conduct will cease.

Oxygen may be cut off suddenly from the tissues by such means as drowning, strangling, or beheading, but the physiologist, psychologist, and doctor will find more to interest them in the effects of partial but prolonged shortage. This generally arises in one of three ways : shortage of oxygen in the air, interference with its passage through the lung membrane into the blood, or failure of the blood to carry it to the tissues. The air breathed may contain too little oxygen if it is diluted with some other gas, or if the oxygen is partially removed from it. On ships, to take a single example, scores of men die every year by entering compartments from which the oxygen has been removed by paint, coal, or grain. On the other hand, the composition of the air may be unaltered, but its pressure reduced ; so that a given volume of air contains a less weight of oxygen than at sea-level. In this case the blood can hold less oxygen, just as soda water can hold less carbon dioxide when we lower the pressure on it. The oxygen in blood is mostly combined with haemoglobin, which gives it its red colour, and the compound formed does not break up appreciably till the pressure of oxygen in the air has been considerably lowered ; so that a small drop in oxygen pressure causes no noticeable effect. The effects of a larger drop have been studied not only in balloons and aeroplanes and on mountains, but by the artificial production of low pressure.

In the factory of an enterprising firm of diving-dress makers in South London is a steel cylinder about seven feet high and five in diameter. It communicates with the outside by a manhole, a small window of very thick glass, and two pipes. With a companion I crawl in through the manhole, which is closed behind us by a formidable series of screws. An engine begins to suck the air out through one of the pipes. The air becomes cold and fills with mist. In five minutes we have reached a pressure of 350 millimetres of mercury, or less than half an atmosphere, corresponding to a height of 22,000 feet above sea-level. I look at the barometer, and open the inlet valve so as to keep the pressure steady. And now I have time to observe my own symptoms. I am breathing rapidly and deeply, and my pulse is at 110 ; but the breathing soon calms down, and I feel much better, though perhaps my writing is a shade wobbly. But why cannot my companion behave himself ? He is making silly jokes and trying to sing. His lips are rather purple, the colour of haemoglobin when uncombined with oxygen. I feel quite unaffected ; in fact, I have just thought of a very funny story. It is true I can't stand without some support. My companion suggests some oxygen from the cylinder which we have with us. To humour him I take a few breaths. The result is startling. The electric light becomes so much brighter that I fear the fuse may melt. The noise of the pumping engine increases fourfold. My note-book, which should have contained records of my pulse-rate, turns out to be filled with the often repeated but seldom legible state-ment that I am feeling *much* better, and remarks about my colleague, of which the least libellous is that he is drunk. I put down the oxygen tube and relapse into a not unpleasant state of mental confusion. An hour later, in spite of our indignant protests, the engine is stopped, and we return to normal pressure, no worse off except for a slight and transitory headache.

For longer experiments a mountain is desirable ; and to avoid the disturbing influences of fatigue on the one hand, and athletic training on the other, it should be ascended by rail. The only railways ascending over 14,000 feet are in the Rockies and the Andes, and it is here that the most complete investigations of prolonged oxygen want have been made. After a few hours nine people out of ten who have ascended rapidly from near sea-level suffer from sickness and headache and may faint. These symptoms are at once cured by a few minutes of oxygen inhalation, and have nothing to do with the low pressure as such. Later a quarrelsome stage generally supervenes. One of the dozen or so permanent residents on the top of Pike's Peak is a sheriff, who is needed to deal with visitors. Later the body begins to adapt itself, and the symptoms pass off more or less completely. The bone marrow manufactures new red blood corpuscles until the blood can hold 20 or 30 per cent. more oxygen than normal when saturated, and slightly more even at a high altitude. The kidney holds back acid which it would normally excrete, and thus goads the respiratory centres in the brain to increased activity. And something seems to happen in the lungs which also occurs in athletic training. The attempts to decide the nature of this change, if any, constitute the most interesting of our inter-'varsity sports. Oxford has pinned its faith to the view that the lung learns to force oxygen into the blood as the gut forces food. Cambridge holds that the gas soaks in as it would through a dead membrane. The contest has now been raging for more than sixteen years, but in spite of American and Danish participation, is still undecided.

As compared with many other mammals, man is very efficient at adaptation. Cats generally die at 14,000 feet, while cows die at 15,000, and give no milk above 13,000, even in the tropics. There is, however, a limit to human adaptability, and it is an open question whether the summit of Mount Everest lies above or below this limit.

The mountaineer has generally time to adapt himself during his approach to the final stages. The airman only spends a few hours a week at most above 10,000 feet, and therefore cannot adapt himself. Moreover, the mountaineer will be warned of his danger by shortness of breath, but the airman will be lured higher and higher by an increasing and unreasonable conviction that he is all right, until he suddenly loses consciousness. When Sully-Prudhomme sang the courage of Sivel and Croce-Spinelli, who died on a balloon ascent in 1875, he was celebrating the psychological effects of oxygen want. Even at 10,000 feet the airman's judgment would probably be improved by oxygen. Above 16,000 feet it is an absolute necessity, and the chief participants in the war supplied the crews of their high-flying aeroplanes and airships with compressed or liquefied oxygen and more or less efficient breathing apparatus.

In various diseases the oxygen cannot pass quickly enough into the blood. Thus, in croup and bronchitis the air passages to the lungs are narrowed, and in lobar pneumonia the membrane through which oxygen passes into the blood is thickened by inflammation. In either case the blood leaves the lungs without its full complement of oxygen. Oxygen has a great future in medicine, and could probably halve the death-rate in pneumonia. But as generally administered it has little more therapeutic value than extreme unction, and is much more expensive. If it is merely blown in the direction of the patient's mouth, he or she does not get enough to soak through the thickened membrane. If his head is enclosed in a box into which oxygen is blown, he rebreathes the carbon dioxide of his expired air and suffers severely. It must be given continuously, sometimes for three days and nights on end. To give it intermittently is like dragging a drowning man to the surface once a minute. It should not be breathed pure, as it is poison, though a rather slow one. Further, the treatment should be started before, and not after, the patient shows signs of approaching death. These conditions are best fulfilled if the gas is administered through a suitable mask, such as that designed for the treatment of war-gas pneumonia. When it is properly administered, the patient's mental state, colour, and other symptoms improve within five minutes.

Finally, the blood may be unable to carry enough oxygen. In heart disease the tissues generally get enough as long as their demand is restricted by keeping the patient at rest. In anaemia and carbon monoxide poisoning a given volume of blood can hold less than its normal amount of oxygen. In the former case the heart increases its output when at rest, but has no reserve of power to fall back on during exercise. In the latter the poison not only displaces some of the oxygen from the haemoglobin, with which it itself forms a compound, but makes the removal of what little is carried unusually difficult, and the brain especially feels the shortage. Carbon monoxide is the poisonous constituent of coal-gas, the gas from charcoal and coke fires, and that produced by explosions in collieries. The abolition of the legal restrictions on the proportion of it in lighting gas is now making the symptoms of poisoning by it familiar.

It might be thought that oxygen would have a good effect on normal people. As a matter of fact, it is poisonous if breathed either at high pressure or for days at a time, while the only effects of non-poisonous doses are to slow down the heart's rate, and, if given for long enough, to diminish the amount of haemoglobin in the blood. The result of these changes is that the tissues get as much oxygen as normally, and no more. Their supply is accurately regulated, and though the physician can sometimes help nature to re-establish the normal state of affairs, he can never improve upon it.

WATER POISONING AND SALT POISONING

THE cells of such an animal as a sponge or a sea-anemone are exposed to fluids of rather variable composition, for the water and salts of their environment penetrate all through their bodies. In man the cells are bathed in the plasma or fluid part of the blood, whose composition is very constant. They are far more efficient at their particular functions than those of a simpler animal, but also far more sensitive to changes in their environment. The mammalian cell may be compared to a civilized man, who, if properly fed, housed, and clothed, is far more efficient than a savage ; but who can only work under a somewhat artificial and narrowly limited set of conditions.

Most of our bodily activities can be regarded as more or less successful attempts to keep our *milieu intérieur* uniform. The lungs, under the direction of the respiratory centres in the brain, regulate its dissolved gases, the kidneys its content of water and salt, the ductless glands its precious burden of substances controlling the rates of cellular oxidation and growth. And much of the activity of the brain and muscles only serves to enable the gut to maintain the normal level of sugar and other foodstuffs.

The effects on the body of a deficiency of foodstuffs or oxygen in the internal environment are well known and fatal ; those of an excess (as in diabetes and oxygen poisoning), though rarer, are equally serious. And the actions of small amounts of various foreign bodies are well known to every student of our criminal proceedings. But it is only of late years that attention has been drawn to the exact regulation of salts and water, which in the main undergo no chemical changes in the body, and to the effects of any departure from their normal proportions.

The story of human water poisoning is as follows. A normal man can get rid of water as fast as he drinks it. A man with severe kidney disease rarely drinks more water than he needs, nor is he encouraged to do so. But the kidneys may be paralysed by other means. An American physician had succeeded, by injecting an extract of the pituitary gland, in temporarily suspending the flow of urine, amounting to some four gallons a day, of a man suffering from diabetes insipidus. But the patient went on drinking water at his normal rate. Rather suddenly he went into convulsions. These were at once relieved by injecting 10 per cent. salt solution into a vein. As the blood plasma in which the corpuscles are carried contains just under 1 per cent. of salt, mainly of the ' common ' kind, this served to bring its water content back to the normal. Experiments on animals made it clear that too high a proportion of water to salt leads to convulsions and death as inevitably as too high a concentration of strychnine. The most probable explanation is perhaps that certain nerve-cells become sodden and swollen, as do those of our skin during a prolonged bath.

Another series of cases appeared in a very different quarter. Perhaps the hottest place in England is about a mile underground in a well-known Lancashire coal-pit, where the miners work in boots and bathing-drawers, and empty the sweat from their boots at lunch. One man sweated eighteen pounds in the course of a shift, and it is probable that even this figure has been exceeded. This sweat contained about an ounce of salt—twice what the average man consumes in all forms per day. The salt loss was instinctively made up above ground by means of bacon, kippers, salted beer, and the like. And as long as they did not drink more than a quart of water underground, no harm came to the miners. But a man who has sweated nearly two gallons is thirsty, and coal-dust dries the throat, so this amount was often exceeded, and the excess occasionally led to appalling attacks of cramp, often in the stomach, but sometimes in the limbs or back. The victims had taken more water than was needed to adjust the salt concentration in their blood, and the diversion of blood from their kidneys to their muscles and skin was so great that they were unable to excrete the excess. The miners in question were offered a solution of salt in water which was of about the composition of sweat, and would be somewhat unappetizing to the average man. They drank it by quarts and asked for more. And now that it has become their regular beverage underground there is no more cramp, and far less fatigue. It is almost certain that the cramp of stokers, and of iron and glass workers, which is known to be due to excessive water-drinking, could be prevented in the same way.

A man who takes an ounce or so of salt without vomiting develops a violent thirst; and if he satisfies it so as to dilute the salt to about the composition of plasma, gets rid of the salt and water rather slowly. Some of the extra fluid accumulates under his skin, and he becomes 'puffy' about the eyes and ankles. In certain types of kidney disease salt excretion becomes very difficult; the patient drinks water to keep the composition of his blood nearly constant, and swells up, developing dropsy. More rarely there is little swelling, but the circulatory system appears to suffer from excess of salt. In either case the patient may be very greatly benefited by adopting a diet practically free from salt. To take a concrete instance; a dropsical patient lost twenty-two pounds when placed on a salt-free *régime*, and regained them within a fortnight when given the ordinary hospital diet. Such dramatic results, obtained by certain French physicians, led to somewhat exaggerated hopes, and all kinds of kidney disease were treated by depriving the patients of salt, regardless of the fact that the diseased kidney may retain its normal capacity for salt excretion while unable to get rid of many much more poisonous substances. Moreover, dropsy is not always due to salt retention.

We are not on such sure ground with regard to most of the other salts in the blood, but it is known, for example, that a certain type of spasms in children is associated with a deficiency of lime salts, and can usually be cured by administering them in sufficiently massive doses. One of our chief difficulties in work in this field is the extreme accuracy of the technique required. The calcium content of the blood is one part in twenty thousand. A competent biochemist will not err more than one per cent. in his estimation of the calcium in ten cubic centimetres of blood, but the analysis requires some hours, and competent biochemists are very rare.

Moreover, so many of the normal constituents of the blood are present in altered amount in disease that it is often almost impossible to distinguish cause from effect. One of the most hopeful lines in modern experimental medicine is the production of small, but definite, chemical changes in the organism, and the careful recording of their effects. Such experiments bear the same relation to clinical observations as did *responsa prudentium* to case law in the Roman legal system, since they enable us to eliminate such features of the disease as are irrelevant to the problem under consideration. The experimental pathologist is apt to miss the less obtrusive symptoms when working on other animals, and in the long run he is driven to use his own body as an instrument of research. And one is certainly amply rewarded when a chemically simple upset of one's composition brings on the main subjective symptoms (from backache to nightmare) of influenza, or the curious facial and manual distortions of tetany.

No doubt only a small proportion of the symptoms of disease will be found to be due to excess or defect of the normal salt constituents of the body. Attempts to explain the causation or effect the cure of cancer on these lines, though often made, are hardly likely to succeed. But the quantitatively minded biochemist will rather interest himself in substances of known composition and measurable concentration than attempt to follow the behaviour of hormones and immune bodies which are as yet neither chemically defined nor accurately measurable. And the principles both of method and of interpretation which develop from the study of so simple a body as common salt may well guide the biochemist of the future in his studies of the physiological and pathological action of substances yet unknown.

IMMUNITY

AS I ate my gorgonzola cheese I attempted to console myself for a violent cold in the head by meditating on some of the ills that human flesh is not heir to.

For that cheese was very ill. It had first been attacked by a gang of most ferocious bacteria and finally by a green mould. But to all its diseases I was immune. If I had eaten a slice of mutton equally full of bacilli they would certainly have proceeded to attack me, while a cheese of milder character would have caught the green malady from the gorgonzola.

Fortunately however man is immune not only to the microbes of cheese but to most of those which attack animals. We do not get distemper from our dogs and very rarely foot-and-mouth disease from our cattle ; though we share with the latter a liability to tubercle, and there is a rare and generally fatal lung disease which old ladies occasionally get from parrots.

We know very little as yet about this natural or inherited immunity. A thousand years hence it will perhaps be the task of medicine to confer it upon all future humanity against all possible diseases. We only know that acquired immunity is not inherited. Although almost all their parents have had measles and thus become immune against a second attack, children not only catch it nowadays as easily as a generation ago but are distinctly more likely to die of it.

When we are not so lucky as to be born immune to a disease we can acquire the immunity in two ways. We can get a mild dose of it or some equivalent malady, or we can take advantage of the acquired immunity of some other man or animal.

The first method, that of active immunization, generally leads to the production of substances in the blood which destroy the microbes or the poisons which they make, and has been most conspicuously successful in the cases of small-pox and typhoid fever.

The former is one of three diseases each of which confers immunity against the others—namely, cow-pox, alastrim, and small-pox. Alastrim is the disease which is now[1] spreading among the unvaccinated in Northern England. Although in its early stages indistinguishable from small-pox, it is not much more dangerous than measles. It never appears to turn into small-pox, any more than does cow-pox.

Unfortunately it is still registered and reported in the newspapers as small-pox. The excuse given for this is that as the two diseases cannot at first be distinguished it is desirable to segregate sufferers from alastrim. But it leads to a very dangerous contempt for the real small-pox, cases of which occasionally arrive in England from abroad and which is just as deadly as ever.

Acquired immunity to these diseases is, of course, not absolute, but fades away in the course of time, though never completely. Mr. Bernard Shaw, for example, developed very mild small-pox in spite of having been vaccinated, and does not let us forget it. The next few years will probably see a revolution in our

[1] 1926.

methods of vaccination, for last year Dr. Gordon showed that animals can be successfully vaccinated with dead cow-pox germs, as men are with dead typhoid bacilli.

Active immunization is quite useless when an acute disease has once started, though it seems to be of some value in chronic infections. It is, however, a most successful prophylactic provided the immunity lasts, as in the case of diphtheria, where the injection of heated or partly neutralized toxin apparently confers life-long protection. It will be one of the tasks of preventive medicine in this century to determine how long such artificial immunities last and against which diseases it is worth while conferring them.

When the illness has once started we can sometimes employ passive immunization, which has had its greatest triumph in the case of diphtheria. Serum from an immunized horse, if given in time, is an almost certain cure. The same is true—on paper—of some types of pneumonia. But while in diphtheria the sore throat gives warning that a poison is being manufactured which may stop the heart in a few days, in pneumonia the lungs are generally damaged when the doctor arrives, though serum might have been invaluable two days before.

Many of the commonest acute diseases of children, such as measles, scarlet fever, and whooping-cough, do not attack animals, so we must get our immune bodies from other human beings. Conspicuous success has attended the injection of serum of convalescents from these diseases in the United States, France, and Germany. Children who had been exposed to infection did not develop the diseases, and those injected in their early stages immediately began to convalesce.

This treatment is now being taken up in London, and we may look forward to the day when the victims of measles may serve to defend their schoolmates instead of infecting them.

BLOOD TRANSFUSION

IT was Sir Christopher Wren, at that time a professor of astronomy, who invented not only the intravenous injection of drugs, but the transfusion of blood, in the year 1659. His success on dogs, wrote Pepys in his *Diary*, ' did give occasion to many pretty wishes, as of the blood of a Quaker to be let into an Archbishop, and such like ; but may, if it takes, be of mighty use to man's health, for the amending of bad blood by borrowing from a better body.'

In 1667, the year after the great fire of London, the experiment was tried before the Royal Society on Arthur Coga, a half-mad Cambridge graduate ' that is poor and a debauched man, that the College have hired for 20s. to have some of the blood of a sheep let into his body. Some think it may have a good effect on him as a frantic man by cooling his blood.' Mr. Coga survived to write an account of his case in Latin, and claimed that he felt a new man, but Pepys records that he continued ' cracked a little in the head.'

As a matter of fact it is doubtful whether he can have received much blood from the sheep ; for subsequent attempts to transfer blood from an animal to man, or from one animal species to another, have generally proved harmful, and often fatal. And though transfusion was occasionally practised with conspicuous success from one human being into another, there were times when the blood even of a near relative acted as a deadly poison

It was not till this century that the reason for these fatalities was investigated. It was then found that very often the red corpuscles of the injected blood are treated in the same way as bacteria to which a man is immune. They are first rendered sticky, or ' agglutinated,' and then broken up, by substances found in the blood serum of the recipient. Fortunately it is not necessary to inject blood in order to test its properties. If we mix a drop of the donor's blood with a drop of serum from the recipient, we can watch what happens with a microscope or even a good hand lens.

Later Jansky and Moss showed that as regards these properties of the blood every human being falls into one of four groups. As a general rule the blood from a member of group 4 can safely be transfused into any one, from group 3 only into groups 1 and 3, from group 2 only into groups 1 and 2 ; while the blood of a member of group 1 is poisonous to all except his own group. Thus the unfortunate members of group 4, to which I belong, can be called on to give blood to any one, while the blood of most other people is poisonous to them.

It was next shown that the group of any individual is fixed from the moment of birth. Finally J. R. Learmonth, a Glasgow medical student, obtained drops of blood from out-patients and their families, and found that membership of a group is inherited in a very simple manner according to Mendel's law. For example if two parents belong to group 4 all their children belong to it also, while a parent in group 1

may have children of any group.   There is no doubt
that this knowledge will be used in the future in many
cases of disputed paternity.

During the war Salonika was an ethnological museum.
Besides the native Greeks, Jews, and Turks, there were
British, French, Italians, Serbs, Russians, Arab, Negro,
Annamite, and Malagasy soldiers, and German,
Austrian, and Bulgarian prisoners.   The brothers
Hirzfeld, who were Polish medical officers in the
Serbian army, obtained samples of blood from some
8000 men and tested them.   They found great differences between different races, and since then tests have
been made all over the world, with most interesting
results.   For example groups 1 and 3 are hardly ever
found among American Indians.   In the old world
they are rarest in British and Belgians, and steadily
increase as one passes towards India or Africa.   The
only coloured race whose groups are in European
proportions are the Japanese.   The proportions are
not modified by environment ; for example they are
the same among pure-blooded Jews in Europe as
among Arabs ; among pure-blooded Gypsies as among
the natives of India, from which their ancestors came.
So it is clear that a careful study of their distribution
will throw new light on the origin of the human races.

While the tailed monkeys and lower animals do not
fit into the human groups, the tail-less apes such as the
orang and chimpanzee, which are our nearest animal
relations, all belong to one or the other of them.

Blood transfusion was first practised on a large scale
during the war.   Since then it has been carried out in
tens of thousands of cases.   Not only can it be used to

combat loss of blood from wounds, childbirth, or operations, but it has saved the lives of many new-born babies suffering from internal haemorrhages, and has at least staved off death for several years in pernicious anaemia. In this country we are still apt to regard blood donors as heroes, though most healthy men can give a quart of blood and return to work next day. But in the United States a more practical point of view prevails, and the professional blood donors have formed a trade union !

CANCER RESEARCH

THE cells of a normal adult divide no more than is necessary for the replacement and repair of the tissues. When there is a local overgrowth the resulting structure is called a tumour. Those tumours, from warts upwards, which show no tendency to spread into surrounding tissues, and have a definite boundary, are called benign, and can generally be removed completely. Malignant tumours or cancers have no such sharp boundary, and are dangerous because they tend to spread, and often to form colonies in distant parts of the body. Researches on cancer fall into two categories, those directed to ascertain its cause, and hence its method of prevention, and those which study its behaviour and the methods of curing or alleviating it.

Our knowledge of the causation of cancer will only be complete when we know why tissue cells do not divide under normal circumstances, though they will often do so if separated from the rest of the body and placed in a suitable nutrient fluid. Meanwhile we may remark that there is no more one cause of cancer than of fever. Cancer may be produced by tar, X-rays, or parasitic worms, just as fever may be produced by a bacillus, a head injury, or a drug. This fact makes it rather unlikely that we shall be able to prevent all forms of cancer by the same method. Every experimental cancer so far produced in animals has been a result of chronic irritation of some kind, and the same is true of the types of cancer affecting various classes of men ; for example, chimney sweeps, tar and paraffin workers, X-ray operators; as well as syphilitic tobacco smokers, who are said to be specially liable to cancer of the tongue. From this fact the generalization has been made that cancer is always due to chronic irritation. Since paraffin cancer may develop ten years after the patient has ceased to work with paraffin, it is clear that the process of induction is slow, and this slowness has been given as the reason why cancer so rarely develops in people under forty years of age. As against this view, it may be remarked that most forms of chronic irritation do not cause cancer, or cancer of the little toe and the nasal mucous membrane would be very common, which they are not. And heredity may account for cancer. Maud Slye, of Chicago, who has done post-mortems on over 40,000 mice, finds that in certain families every mouse living over eighteen months develops cancer ; in others, kept under similar conditions, none ; whilst the disease is inherited, approximately at least, as a Mendelian recessive, as it undoubtedly is in certain flies. However, tar will cause cancer in mice of any family.

Human statistics give no indubitable proof of heredity in this case, for a slight tendency to run in families, such as certainly exists, may be explained by similar environment or habits. But it is likely that susceptibility to cancer, as to other diseases, is partly inherited. Unfortunately, as the malady generally appears after the age of bearing or begetting children is past, natural selection does not weed out those who are liable to it, as it is weeding out those liable to tuberculosis, which commonly kills its victims while they are still potential parents, or earlier.

As cancer is far commoner in civilized than in savage communities, it has been attributed by various authors to most of the conditions peculiar to civilized life. To take the latest example, Mr. Ellis Barker has attributed it to food preservatives, vitamin deficiency, and chronic constipation. Any or all of these may be true causes, but there is no evidence to associate cancer with rickets, scurvy, beri-beri, or any other vitamin-deficiency disease in man, and out of the thousands of animals which have been starved of vitamins only one, so far as I know, has developed cancer. There is not a jot of evidence that food preservatives cause it, while these substances certainly stop a good deal of bacterial infection. With regard to constipation, we still await either statistical or experimental data to connect it with cancer.

If I were compelled (which fortunately I am not) to suggest a cause of cancer among the features of civilization, I should point to the use of coal, which when burnt yields the cancer-producing body found in tar and soot. There is some rather inadequate statistical evidence from Scotland that the use of peat in place of coal fires is associated with immunity to the disease. I should look askance at the indiscriminate use of such substances as vaseline, which may very well contain the cancer-producing body found in paraffin, and lubricating oil, which certainly contains it. The tar used for road-making will in all probability cause cancer in many of those who spread it, conceivably in those who inhale it in the form of dust. Such possibilities as to the cause of cancer have at least a solid ground in occupational mortality statistics and in experimental work. Speculations as to diet have not, except as regards the action of one or two dyestuffs which are rarely if ever found in human diet, though they produce internal cancer in those who make them.

When we turn to the study of the cancer cell and its habits, our most practically valuable knowledge is as to its appearance under the microscope. This often enables us to decide whether a given growth is malignant or not, and hence whether it is the surgeon's duty not merely to remove the growth of which a microscopical sample has been taken, but to cut away neighbouring tissues into which its cells may have emigrated. The large amount of work which has been done on the microscopical structure of appropriately stained slices of cancers has therefore saved many lives, and removed from many more minds the ghastly suspicion that a harmless growth might be malignant. Almost equally useful has been the work of the pathological anatomists, who by innumerable post-mortem examinations have discovered the paths along which cancer cells might migrate from certain sites. The most usual of these paths are the lymphatics, vessels which drain a clear fluid from the inter-cellular spaces of the tissues into the veins. On the way to the blood-stream it passes through lymph nodes, in which foreign bodies such as dust particles, bacteria, and cancer cells are stopped. Thus, if a cancer of the breast spreads, it will generally spread along the lymphatics which pass through the nodes of the armpit. When a breast is removed for cancer, the lymphatics which drain it are also dissected

out for some distance.   If the lymph nodes of the armpit are free from cancer cells the patient has a good
chance of surviving for many years, though, unfortunately, migration sometimes takes place into the
inside of the chest and belly.   The same principles
apply to other sites.   Nearly six years ago a friend of
mine developed cancer of the large intestine.   Two
feet of that organ were removed, and a certain length of
lymphatics draining it.   On each of the possible routes
of migration at least one lymph node was found free
of cancer cells.   My friend is alive and well to-day.

As to the more intimate nature of the cancer cell,
we know but little.   Our most important information
as to the differences between it and the ordinary cell
has come in the last few months from the laboratory of
Warburg in Berlin.   There are two ways in which a
cell may use sugar as a source of energy.   It can burn
it, that is to say, combine it with oxygen, or without
using up any oxygen it can split it into simpler products.   Thus a yeast cell splits it into alcohol and
carbon dioxide, a contracting muscle fibre into lactic
acid.   If the muscle cell has a sufficient supply of
oxygen, it soon puts the split sugar together again,
burning some of it to get the required energy.   About
eleven times as much energy can be got from sugar by
burning it as by splitting it into lactic acid.   But the
cancer cells, even in presence of much oxygen, will split
far more sugar than they oxidize, and in this they differ
from all other cells, except those of very early embryos.
They are, in fact, spendthrifts, relying on the normal
cells of the body for constant supplies of sugar.   How
to exploit this knowledge is at present beyond us.   The

biochemist is here in the position of a detective who is watching a suspected person, and has just observed in him some very peculiar behaviour, not, however, peculiar enough to bring him definitely within the grasp of the criminal law. But there are numerous possibilities. So many curious facts about the chemical behaviour in the body of sugars and their likes have turned up, especially in the course of investigations on diabetes and on muscular exercise, that we may easily discover a method of discouraging the cancerous type of sugar metabolism without affecting that of a normal cell.

Among the mass of published work on cancer almost all the data necessary for prevention or cure may already exist. The majority of this work is worthless. For in cancer research it is legitimate to follow up a line of investigation even if one feels that the chances are a thousand to one against its success. But after a year or two's work this detached point of view as to one's own importance may be succeeded by one more resembling that of the average person. One does not like to dismiss a year's research in ten lines of print. And the importance of the end in view may lead to an emotional condition which is fatal to scientific thinking. An objective attitude is almost as difficult as with regard to spiritualism, alcoholism, or birth control. But in the study of cancer, as elsewhere, the one hope of humanity lies in the adoption of that attitude.

NOTE.—Since this essay was written in January 1925 a good deal more has been found out about cancer, particularly with regard to its transmissibility by cell-free filtrates. The facts given above appear, however, to be unaffected by subsequent research.

THE FIGHT WITH TUBERCULOSIS

TUBERCULOSIS does not stand first on the list of causes of death in England, but it is the most serious, because it kills in infancy and prime of life, whereas cancer is a disease of the old, and most other diseases affect them more than the young. Although the death-rate from tuberculosis has been halved in the last twenty-seven years, yet it has not been reduced to the condition of a rarity like typhoid fever, or a complaint from which one only dies by one's own neglect or that of one's parents and guardians, like small-pox.

A great number of attempts have been made of late to cure it by some method more direct than good feeding, fresh air, and sunlight or ultra-violet radiation. But the tubercle bacillus is a tough creature, and it is hard to kill it without killing its host first. Sanocrysin, the gold compound advocated by Mollgaard, of Copenhagen, has proved effective in some cases, but it has caused kidney trouble and fever in others. A great many attempts have been made to cure tubercular disease by increasing the immunity of the patient to it in various ways. Most of these attempts have been failures.

At the present time the Spahlinger treatment, which belongs to this class, is under investigation. It has not received a very thorough trial, because Dr. Spahlinger, unlike all other workers who have investigated the question from a scientific standpoint, has not seen fit to publish full details of his method. And the value of any treatment can only be estimated when it has been applied to some hundreds of cases.

It is worth remarking, however, that success on these lines seems to be much less hopeful than in the case of acute diseases such as scarlet fever or measles, a single attack of which generally confers immunity in future.

The most interesting attempt to confer immunity is one which aims at prevention and lays no claim whatever to cure. Professor Calmette, sub-director of the Pasteur Institute in Paris and a Fellow of the Royal Society, who has been engaged on the problem for over twenty years, believes that he is able to render most babies immune to tubercle for life.

For thirteen years tubercle bacilli were grown on bile-soaked potatoes in his laboratory. At the end of this time they had become domesticated, so to speak, and although they multiplied when injected into animals they no longer caused them illness. Over 3000 new-born calves were injected in this way, and it was found that they had become immune to ordinary tuberculosis.

After experiments on monkeys in West Africa the treatment was applied to new-born babies four years ago. Since July 1924 over 5000 French infants have been given doses of the ' bacille Calmette-Guérin,' as the tame tubercle germ is called. Six hundred of these were children of tuberculous parents, of whom one in four usually die in their first year. Instead of 150 only 11 actually died. The results among the children exposed to less intense infection were, of course, better.

Adult Europeans are generally already slightly infected with tubercle, and infection with the new bacillus is useless or dangerous. Negroes, however, are little attacked until they come to a temperate climate, where they die of tuberculosis in great numbers. Accordingly, in an experiment now in progress one half of a Senegalese battalion in France have been inoculated, and their health is being compared with that of the other half.

As one quarter of the deaths of French children in their first two years are due to tubercle, and a large proportion of the French troops are coloured, Professor Calmette's work, if successful, will do more for the future security of France than any number of treaties or pacts.

Outside France similar protection has been afforded to thousands of coloured children in the French colonies. In another five years it should be possible to pass a definite judgment on its value.

But meanwhile the known possibilities of prevention in this country might well be exploited. The greatest single channel of infection is milk from tuberculous cows drunk in infancy or early childhood. But the vast majority even of well-to-do parents do not take the trouble to obtain Grade A or Grade A certified milk for their younger children. In many places it is, of course, not available, but it would be if an economic demand for it existed. And with no public opinion behind it in the matter the Government cannot be expected to legislate drastically in favour of pure milk.

If science has not discovered a cure or an infallible preventive for tuberculosis, it has at least shown how the mortality could be greatly lowered. For the price of a cigar or a cinema a week you can protect your child against its most dangerous enemy. Is it worth while ?

FOOD POISONING

WITH the reappearance of the sun [1] has come this year's first serious outbreak of so-called food poisoning. The great majority of cases occur during warm weather, and this summer we may expect an unusually large number owing to the coal stoppage. For the most potent safeguards that we have against this danger are the kitchen range and the gas cooker.

England is fortunate in producing very few naturally poisonous foods. Occasionally a poisonous fungus is found among a batch of mushrooms, and in 1921 three ladies in Liverpool nearly died from eating mutton stuffed with sage which contained belladonna leaves. But in Japan nearly a hundred people die every year from eating a poisonous fish, the fugu ; and Cuba boasts of no less than seventy-two different species of fish which may cause death or illness.

This country has seen a few outbreaks of arsenic, zinc, and copper poisoning, mostly in beer or cider, including the terrible catastrophe in 1900 which killed at least seventy people, and affected nearly ten thousand. But in more than nine cases out of ten the poison is made by bacteria, much the commonest source being an organism called *Bacillus aertrycke*. Many bacteria manufacture toxins, but fortunately these are generally destroyed by boiling or by the digestive juice. Diph-

[1] May 1926.

theria and tetanus toxins are deadly if injected, but, like snake venoms, they are harmless when taken by the mouth or injected after boiling. The bacteria which cause ordinary putrefaction are almost harmless to adults when taken by the mouth, and though if allowed to act on food for a sufficiently long time they produce moderately poisonous substances called ptomaines, it would require a very heroic bacteriologist to eat meat rotten enough to contain a fatal or even dangerous dose of them. In spite of the verdicts at several inquests it is extremely unlikely that any one has ever died of ptomaine poisoning.

The *Bacillus aertrycke* and its allies, however, make a poison which resists both cooking and digestion. If the contaminated food has been cooked after its infection the bacteria are dead but the poison remains. When the food is eaten it causes intense vomiting and other symptoms of digestive upset, but very rarely death. If the living germs are swallowed they grow for some time in the patient and may continue to manufacture their poisons for a week or more. Death is a good deal commoner in these cases, but even here the vast majority recover completely.

The source of the poison is tinned food in about half the cases, and it is particularly interesting that fifteen out of the last sixteen tins of meat causing poisoning in England came from South America, although only about half of our tinned meat comes from that continent. Potted meat, meat pies, and other products made from scraps of meat come next on the list, but milk, ice cream, and cheese are also occasionally poisonous. Vegetables and fresh meat are much more rarely dangerous. In almost all cases the food has appeared to be perfectly sound, and given no warning either by taste or smell.

How do the bacteria reach the food ? It is almost always infected after cooking, and the bacteria are then given time to grow before it is eaten. Pigs, mice, and rats seem to be the most usual sources of infection, though occasionally human beings and cattle have been incriminated. In many cases the poisonous meat had been kept under filthy conditions. Thirty-seven people were poisoned at Derby in 1921 by pork pies made in a room adjoining a slaughter-house. Nearly a thousand cases of illness and one death were caused at High Wycombe in 1923 by ice-cream made in a disused stable overlooking a yard full of refuse of all kinds. Much food poisoning could undoubtedly be prevented if the manufacture of ice-cream and ' made-up ' meat foods were only permitted in licensed and properly inspected places.

The most serious kind of food poisoning is botulism, of which only one outbreak has been recorded in Britain. Eight people ate ' wild duck ' paste sand-wiches at Loch Maree in 1922, and all died from a paralysis which began in the muscles of their eyes and spread until they were unable to breathe. The poison is made by a bacillus which can only live in the absence of oxygen. It is therefore mainly found in tinned or bottled foods, but occasionally in the interior of sausages and hams. The toxin of *Bacillus botulinus* is the most poisonous of all known substances when taken by the mouth. About sixty pounds of it would probably be sufficient to kill the entire human race.

Fortunately it is destroyed by cooking. In America most of the hundreds of cases on record have been caused by preserved vegetables, such as string beans, spinach, olives, and asparagus, and in view of the large amounts of American canned vegetables imported into England the occurrence of an outbreak from this source is only a question of time.

To sum up, a great deal can and should be done to check food contamination at its sources, but nothing but some minutes' insertion in boiling water can render food in tin, stoneware, or glass containers absolutely safe for human consumption.

THE TIME FACTOR IN MEDICINE

THE average man or woman goes to the doctor to be cured of some disease or injury, and for this purpose expects either surgical treatment or something out of a bottle. And the critics of modern medicine complain that while the surgical treatment is often unduly violent, the medicine is usually ineffective except as a generator of faith. They also point out that in medical teaching enormously greater stress is laid on diagnosis than on treatment. Fortunately for the medical profession, its critics commonly support some therapeutic system such as faith-healing, osteopathy, or herbalism, which is quite demonstrably less efficient than that of ordinary medicine. Again, the study of immunity has fallen into some disrepute because, although immune sera are often potent prophylactics, they are not of much value in curing diseases other than diphtheria.

The reason for this is a simple one. The doctor is generally called in to cure a scar. Diphtheria and one of the types of pneumonia are both bacterial diseases which can be cured by injections of serum. In the former case death is generally due to poisoning of the heart by toxins made by bacteria in the throat. A sore throat is one of the earliest symptoms of the disease, and its typical appearance generally enables a doctor, provided he is called in in time, to inject antitoxin which will save the heart. In pneumonia, on the other hand, the patient generally dies because oxygen cannot enter the blood through the inflamed and thickened lung membrane. By the time the doctor is sure of his diagnosis the inflammation is usually already severe, and most of the pneumococci are already dead. It is merelyvindictive to kill the rest with an immune serum; the problem is to keep the patient alive until his or her lungs recover their normal permeability. Here then is a case where therapeutics fail, not because of our lack of therapeutic methods, but because diagnosis is not yet sufficiently advanced. And it is quite typical. Most cancers can be cured by a sufficiently early operation. It is only in the case of the most obvious sites, such as the breast and tongue, that a diagnosis is commonly made before the new growth has spread so far as to be ineradicable. Elsewhere the doctor is often in doubt until it is too late. Perhaps one middle-aged man in ten with chronic abdominal pain is developing cancer, and his life would be greatly prolonged by operative treatment, but one cannot open up the other nine to make sure.

This is why a serological or chemical test for cancer (of which several have been described, but none universally adopted) would be of more value than almost any advance in treatment which seems immediately probable.

But early diagnosis of disease is the business of the general public even more than of the medical profession. To take an obvious case, venereal diseases in their very early stages are easily and rapidly curable, but every day's delay renders the case slower and less certain. If this fact, and the early symptoms of these diseases, were thoroughly and universally known, hundreds of millions of years of human suffering could be immediately prevented.   But it is public opinion, and certainly not medical obscurantism, which makes a dissemination of such knowledge impracticable at present.   It could not be done without a gross violation of the law relating to indecency.   A few local authorities have attempted to call attention by public advertisement to the early symptoms of cancer of the womb, but their example has not been followed.

As a matter of fact, it would be psychologically unsound to disseminate too wide a knowledge of disease among a people who are ignorant of the working of their bodies in health.   The study of medicine by laymen is said, probably with truth, to conduce to the spread of imaginary maladies.   The study of physiology generally leads to a healthy and amused interest in the normal working of one's own body, against which background the early symptoms of disease stand out. Unfortunately, however, physiology is hardly taught except to medical students.   Its place is occasionally taken in school curricula by ' hygiene,' which usually takes the form of inculcation of ' scientific ' rationalizations of the current views on cleanliness, exercise, and abstinence.   The attempts which are made in such courses to make as many physiological phenomena as possible point a moral, and to suppress the rest, are reminiscent of the analogous attempts to moralize zoology which were made by the authors of mediaeval bestiaries.   Fortunately most children find ' hygiene ' very dull, so less harm is done than might be expected.

But in addition to prevention and cure, the doctor

has another very important duty to his patients, namely, prognosis. Indeed, this is the main duty of doctors working for insurance companies. And in prognosis time is an all-important element. A man of fifty goes to the doctor with heart trouble. He is quite aware that he is likely to die earlier than a healthy man. But it makes all the difference in the world to him whether his expectation of life is two or twenty years. And a doctor may be able to give him this information, even when, except for warning his patient against over-exertion, he can do nothing to check the progress of the disease. The rules governing such prognosis are often rather simple, and are due to the late Sir James Mackenzie as much as to any one man. As a general practitioner at Blackburn he was able to observe the progress of disease in individual patients for periods of over thirty years, and to form a far better opinion as to the prognostic value of given symptoms than a specialist or hospital physician. With some types of heart disease symptoms of a fairly well-defined intensity mark the approximate half-way stage between onset and death. If the patient with these symptoms had rheumatic fever twenty years ago, he has probably another twenty years to live ; but if the causative infection was of quite recent date, he will be well advised to set his affairs in order and live each day as if it were his last.

ON BEING ONE'S OWN RABBIT
THE STORY OF A SKIRMISH IN THE WAR ON DISEASE

MOST educated people have a rough but fairly accurate idea of the methods employed by the bacteriologist in fighting disease. But in many cases we cannot deal directly with the invading organisms, or they have already done irreparable damage by the time the first symptoms appear. Often, too, the cause of sickness is an unusual demand on the body's resources, such as pregnancy or the very rapid growth of babies; and under such conditions constitutional weakness or unsatisfactory diet may lead to serious results.

In all these cases we need a knowledge of how the body works, and how to supplement its resources. If the kidney has been damaged we can often put the patient on a diet which gives it so little work that it can still carry out its functions. If the part of the pancreas which makes insulin has been destroyed, we give the patient daily injections of insulin from the pancreas of pigs. And very often, if we can only relieve the symptoms and keep the patient healthy, the body exercises its marvellous capacity for recovery. The surgeon puts a broken leg in splints. The biochemist provides, so to say, chemical splints for damaged organs.

The story that I have to tell is of the discovery of an improved method of treatment for a rather unimportant disease, a discovery in which I happened to play a part. It began, like most scientific work, with the investigation of a very abstract problem, and the original workers had no idea whatever of how their results would be applied to practical medicine.

I came into the story with no humanitarian motives. I wanted to find out what happened to a man when one made him more acid or more alkaline. The chemists told me that my body was a system of negatively charged colloids. They also told me that when one makes such a system more alkaline the electrical charge on the colloids increases, and that when one makes it more acid it diminishes. But they had apparently never wondered what a colloidal system felt like when one diminished its charge.

One might, of course, have tried experiments on a rabbit first, and some work had been done along those lines ; but it is difficult to be sure how a rabbit feels at any time. Indeed, many rabbits make no serious attempt to co-operate with one. I except always a large buck called Boanerges (which is, being interpreted, the Son of Thunder). Boanerges had to breathe carbon monoxide every day. He sat on the table with his nose in a well-greased funnel. When he got bored he stamped. That was before the war, so no doubt the noise impressed me more than it would now, but I seem to remember that any glass one left on the table collapsed into rather fine dust. If one took no notice of his first stamp he proceeded to walk off. However, he was always willing to co-operate for such a period as he thought reasonable ; but most rabbits get frightened, and to do the sort of things to a dog that one does to the average medical student requires a licence signed in triplicate by two archbishops, as far as I can remember.

A human colleague and I therefore began experiments on one another. Before relating what happened in these tests, it may be as well to discuss briefly the chemical facts with which we had to deal. Acid substances are acid because when dissolved in water they break up so as to yield hydrogen ions, that is to say, atoms of hydrogen which have lost their one electron, and thus acquired a positive charge. Pure water contains a few hydrogen ions—to be accurate, one part in ten thousand million by weight—and the concentration of hydrogen ions is greater than this in acid solutions, less in alkaline. These small concentrations cannot be measured directly, but are estimated by conductivity measurements, by the electro-motive force developed by hydrogen going into solution from a platinum electrode, or by means of colour changes in certain organic substances.

The importance of hydrogen ion concentrations near neutrality was first realized by biochemists. It was found that the hydrogen ion concentration in the human blood was extraordinarily steady, being just on the alkaline side of neutrality. In fact, except for occasional abnormal people, it is doubtful if any variations at all from the usual value have been observed in healthy human beings at rest. The most alkaline healthy blood on record belonged to a conscientious objector ! Each tissue seems to have its own normal hydrogen ion concentration. As soon as the constancy of these concentrations was discovered it became of interest to see, firstly how they were kept steady, and secondly what happened if they went wrong. The two are really bound up together, because among the most striking effects of an upset are the body's efforts to remedy it.

It was learned that the most rapid means of regulating neutrality was through the breathing. The lungs supply the body with oxygen, and remove the carbon dioxide formed by the oxidation of food. The breathing is not regulated by need for oxygen, for a small decrease in the oxygen of the air breathed does not increase the ventilation of the lungs appreciably, nor does any increase slow it down. This is because the blood leaving the lungs is already almost saturated with oxygen, and an increase in the lung ventilation gets hardly any more in, nor does a small decrease appreciably lower the uptake. The chief effects of a changed rate of breathing are on the amount of carbon dioxide (or carbonic acid) lost per minute, and it is the amount of carbon dioxide in the blood which normally regulates the breathing.

The kidneys also help to keep the tissues neutral by excreting excess of acid or alkali, but their action is far slower. It is their function to remove from the blood which passes through them substances which are foreign to it, or which are present in excess of certain standard amounts. Human blood is generally a little too acid, as the sulphur and phosphorus of our foodstuffs are oxidized to sulphuric and phosphoric acids in the body. All the former and about half the latter are excreted in combination with ammonia, which is formed in the kidneys as required to neutralize the acids.

One of the experiments designed to show that carbon dioxide is the normal regulator of breathing was as follows. The subject breathed as fast and deep as he could for two or three minutes. After this he had no desire to breathe for some time, until in fact most of the carbon dioxide blown out had accumulated again owing to the constant steady oxidation in the tissues. By this time he was often blue in the face with oxygen want. These experiments were made to test the method by which breathing is regulated, and those who carried them out were more worried than interested by certain extra effects which they noticed. After about half a minute they got violent ' pins and needles ' in the hands, feet, and face, and after three or four their hands became curiously stiff, and sometimes their wrists bent involuntarily.

In 1920 Collip (who afterwards co-operated with Banting in the isolation of insulin) and Backus, in Alberta, Canada, noticed that the symptoms produced by forced breathing were largely those of slight tetany. Tetany, which must not be confused with tetanus, is a disease characterized by a cramp of the hands, feet, face, and sometimes the windpipe. It occurs in babies (generally in conjunction with rickets) ; in pregnant women ; in adults whose parathyroid glands (four bodies in the neck, each about the size of a pea) have been injured, removed, or diseased ; in diseases characterized by chronic vomiting, and sometimes for no obvious cause. It is much commoner in Germany and Austria than in this country. This was so even before the war, but since then it has greatly increased among children in Central Europe, owing to their unsatisfactory diet.

At about the same time Grant and Goldman, of Washington University, breathed harder and longer than Collip, and obtained all the symptoms of tetany.

Poor Goldman on one occasion, after about half an hour, uttered a shrill cry and went into a general convulsion. Every muscle in his body was contracted, his limbs stretched out stiff, and his back arched.

I have never had a general convulsion as a result of self-experiments in over-breathing. My star turn is probably intense sweating, which breaks out after about twenty minutes. I also probably hold the endurance record of one and a half hours' continuous spasm of the hands and face, though on that occasion I never breathed so hard as to cause the cramp to spread above the elbow, as it does in severe tetany.

The chief trouble in a long experiment is that one tends to drop asleep and stop breathing, so a ruthless colleague is needed to prod one. Perhaps the oddest thing about such spasms is that they leave no bad after-effects, though it is true that certain signs of increased irritability of the nerves may persist for a fortnight.

In our experiments on the effects of acids and alkalis on the human body, my colleague Dr. H. W. Davies and I made ourselves alkaline by over-breathing and by eating anything up to three ounces of bicarbonate of soda. We made ourselves acid by sitting in an airtight room with between six and seven per cent. of carbon dioxide in the air. This makes one breathe as if one had just completed a boat-race, and also gives one a rather violent headache. We analysed large amounts of blood and urine, and found out roughly what changes were occurring in them.

But we still wanted something which would keep one acid for days at a time. Two hours was as long as any one wanted to stay in the carbon dioxide, even if the gas chamber at our disposal had not retained an ineradicable odour of ' yellow cross gas ' from some wartime experiments, which made one weep gently every time one entered it.   The most obvious thing to try was drinking hydrochloric acid.   If one takes it strong it dissolves one's teeth and burns one's throat, whereas I wanted to let it diffuse gently all through my body.   The strongest I ever cared to drink it was about one part of the commercial strong acid in a hundred of water, but a pint of that was enough for me, as it irritated my throat and stomach, while my calculations showed that I needed a gallon and a half to get the effect I wanted.

I therefore had to think of a dodge for getting the hydrochloric acid in under false pretences.   If one gives this acid to an animal it is not got rid of as such by the kidneys, as it would corrode the urinary passages ; but about two-thirds (though not all) is neutralized by ammonia made in the body, and excreted as ammonium chloride.   The same thing occurred in my own case. Now, in the chemical laboratory, when a reaction does not go all the way, it generally means that it can be reversed.   For example, lime and chlorine dissolved in water combine to make chloride of lime, but there is always a little chlorine left over, which one can smell, and conversely one has only to dissolve chloride of lime in water for chlorine to be given off.   So here I argued that if one ate ammonium chloride, it would partly break up in the body, liberating hydrochloric acid.

This proved to be correct.   As a matter of fact, the ammonium salts are poisonous when injected into the blood stream, and the liver turns ammonia into a harmless substance called urea before it reaches the heart and brain on absorption from the gut. The hydrochloric acid is left behind and combines with sodium bicarbonate, which exists in all the tissues, producing sodium chloride and carbon dioxide. I have had this gas produced in me in this way at the rate of six quarts an hour (though not for an hour on end at that rate). Possibly my liver, had I been able to see it, would have resembled a Seidlitz powder, but even had I had a window through which to watch the process I should have been too busy breathing to pay much attention.

Not merely, however, has one to get rid of the carbon dioxide made in the liver, but, in order to preserve the hydrogen ion concentration of the blood as near its normal level as possible, one has to keep the amount of carbon dioxide in it at half or less than half the normal amount, thus compensating for the acidity caused by the hydrochloric acid. It is all very well to breathe four times the normal volume of air per minute when sitting in a chair, but this is a very different proposition when one is walking, and such exercise as cycling becomes quite impossible. I was able to take ammonium chloride at the rate of about an ounce a day for two or three days, and then remained breathless for another two or three, by the end of which time my kidneys had got rid of most of the liberated acid.

I was quite satisfied to have reproduced in myself the type of shortness of breath which occurs in the terminal stages of kidney disease and diabetes. This had long been known to be due to acid poisoning, but in each case the acid poisoning is complicated by other

chemical abnormalities, and it had been rather uncertain which of the symptoms were due to the acid
as such. Moreover, a number of unexpected and
interesting effects occurred. For example, my blood
lost about ten per cent. of its volume, my weight
dropped seven pounds in three days, while my liver,
perhaps as a protest against being treated as a Seidlitz
powder, refused to store sugar, which is one of its
normal functions.

The scene now shifts to Heidelberg, where Freuden-
berg and György were studying tetany in babies.
They had read Grant and Goldman's work, and given
themselves tetany. And although in most cases of
tetany the blood is no more alkaline than usual, it
occurred to them that it would be well worth trying
the effect of making the body unusually acid. For
tetany had occasionally been observed in patients who
had been treated for other complaints by very large
doses of sodium bicarbonate, or had lost large amounts
of hydrochloric acid by constant vomiting ; and if
alkalinity of the tissues will produce tetany, acidity may
be expected to cure it. Unfortunately, one could
hardly try to cure a dying baby by shutting it up in a
room full of carbonic acid, and still less would one give
it hydrochloric acid to drink ; so nothing had come of
their idea, and they were using lime salts, which are
not very easily absorbed, and which upset the digestion,
but certainly benefit many cases of tetany.

However, the moment they read my paper on the
effects of ammonium chloride, they began giving it to
babies, and were delighted to find that the tetany
cleared up in a few hours. Since then it has been used

with effect both in England and America, both on children and adults. It does not remove the cause, but it brings the patient into a condition from which he has a very fair chance of recovering. As a matter of fact, children generally recover in the course of a fortnight or so when treated with cod liver oil; but one cannot wait a fortnight when the child's face and limbs are contorted, and its breathing interfered with. Later on Collip returned to the problem and obtained a substance from the parathyroid glands which will cure tetany in adults, but is said to be less effective in children.

The above episode is quite typical of modern biochemical investigation. An immense number of experiments are being done on human beings, especially perhaps in the United States. For rough experiments one uses an animal, and it is really only when accurate observations are needed that a human being is preferable. For example, the discovery of insulin, which abolishes the symptoms of diabetes, was only possible by experiments on animals, for the simple reason that one cannot inject large amounts of substances of unknown composition into men to see what will turn up. But the elucidation of its mode of action is coming in large part from experiments on human beings.

It might be thought that experiments such as I have described were dangerous. This is not the case if they are done with intelligence. Naturally one only drinks or breathes substances whose probable effects are fairly well understood, and which are known not to be fatal to animals in small quantities. One works up only gradually to the size of dose which produces really striking symptoms. Experiments in which one stakes one's life on the correctness of one's biochemistry are far safer than those of an aeroplane designer who is prepared to fall a thousand feet if his aerodynamics are incorrect. They are also perhaps more likely to be of benefit to humanity in general.

Again, biochemistry, like all science, is strikingly international. Two Germans synthesized Grant and Goldman's idea with my own, and I am now working at the neurological side of tetany in Paris during my vacations in conjunction with a French physiologist. And recently I read the confirmation, by a worker in Moscow, of some of my work on inheritance in poultry.

It is on such lines as the above that medicine is advancing most rapidly at the present moment. Pasteur's discovery of the microbial origin of infectious diseases and the subsequent work on immunity to them led to immense advances in preventive medicine. Water-borne diseases such as typhoid and cholera have been abolished in civilized countries. Insect-borne diseases, such as plague, malaria, and infantile diarrhœa, could also be abolished if people seriously wished to be rid of them. So might venereal diseases. But when a person is once ill, there are few complaints which can really be dealt with successfully on Pasteur's lines.

Chemical methods of cure have been more satisfactory. Sometimes we use a definite chemical substance to destroy the parasite without harming the patient, as in Ehrlich's cure of syphilis ; sometimes, as in the case which I have described, we concentrate on relieving the patient's symptoms, and hope that if he is kept alive he will overcome the microbic invasion or nutritional upset that has caused the disease. A striking example of this method is the treatment of pneumonia by continuous inhalation of oxygen. Most deaths in lobar pneumonia occur because the lung is so thickened that oxygen cannot get through it to the blood unless there is very much more oxygen than usual in the air breathed. These cases do not die if they are given air rich in oxygen for three or four days on end.

But if methods of this type are to be employed with success in medicine, we shall have to make considerable demands on the intelligence, accuracy, and honesty of every one concerned—doctor, nurse, pharmacist, and patient. For example, it is perfectly safe to take two-thirds of the dose of ammonium chloride which would kill one by liberating enough hydrochloric acid to combine with all the alkali in one's body. And one gets rather little effect from anything under one-third of the lethal dose. The same is true of oxygen. The air breathed by the pneumonia patient should contain at least half its volume of oxygen ; on the other hand, pure or nearly pure oxygen will probably kill him in forty-eight hours !

Accuracy of the kind needed is perfectly attainable. We entrust our lives every day with complete confidence to the accuracy of engineers, railway signalmen, and omnibus drivers. When people realize that biology is as exact a science as physics, and that medicine will one day be as exact an art as engineering, we may hope for some real progress in the cure of disease. But at present a doctor knows very well that his patient is likely to forget his medicine on Saturday and take a double dose on Sunday. He dare not put anything in the bottle that is likely to kill, and in consequence there is very often nothing likely to cure either. In the same way certain substances when injected are fatal if the patient has been injected with them before. Until the doctor can rely on the patient neither lying about his or her former medical history nor forgetting it, he will be very chary about using some of his most effective remedies. Civilized life demands intelligence and education, not in a certain class only, but in the whole community.

The medical profession have perhaps not always done all they might in educating the public in the facts of their science. No doubt this is partly due to a survival of the ' medicine man ' tradition, but another reason is that a half-educated patient who tries to diagnose his own disease is often worse than a completely uneducated one. Before people tell the doctor that they are suffering from heart disease, they should realize that a pain felt in the region of the heart is most commonly due to irritation of the stomach ! But even so, a slight knowledge of the facts of medicine is becoming essential, not only if patients are to co-operate with their doctors, but if such diseases as cancer are to be recognized and dealt with before they have gone too far to be curable.

Finally, since the public has begun to pay for medical research, it has a perfect right to know how its money is being spent. During last year about one part in four million of the national revenue was employed during some weeks in keeping me awake during attacks of tetany, and in analysing blood samples drawn from me in the course of them. It has been the object of this article to suggest that one-four-millionth of the nation's income was well spent.

WHAT USE IS ASTRONOMY ?

THERE has been an Astronomer Royal for two hundred and fifty years, but there is no Physicist Royal nor Bacteriologist Royal, although during the last fifty years physics and bacteriology have been of greater service to the State than astronomy. And the taxpayer may sometimes be tempted to ask what return he gets for the money spent on Greenwich Observatory. There cannot be the faintest doubt of its value during its first two centuries of existence. Navigators depended on observations of the sun, moon, and stars to a far greater extent than now. There were no lighthouses to give them their position, no accurate charts, no wireless, and above all, a sailing ship was vastly more likely than a steamer to deviate from its intended course. Accurate astronomical tables were not only required for the purposes which they now serve ; but until Harrison invented the chronometer, the only satisfactory method of obtaining the time at sea was by observing the occultation or covering of stars by the moon or of his satellites by Jupiter. And so Greenwich Observatory played a very important part in the foundation of the British Empire.

But the nautical almanac could now be kept up to date (or rather three years ahead) by a few calculators whose results were checked by a single telescope ; and the large majority of astronomers now interest themselves not so much in the motions of the sun, moon, and planets, as in the distances, composition, and temperatures of the fixed stars, or in the structure of the sun, and their observations are certainly of no use to navigators.

But that is not to say that the Astronomer Royal is not earning his salary. For the greatest benefits of astronomy have been indirect and unperceived. I fear that few racegoers as they take out their field-glasses bless the name of Galileo, who made the first at all powerful telescope in order to observe the stars. Nor does the engineer or surveyor always remember that both trigonometry and logarithms were invented by astronomers to aid them in their calculations. Again, common sense tells us that we see things as they are. It was an astronomer who, by observing that the eclipses of Jupiter's moons were later than theory demanded when they were farther away from the earth, showed that we see things as they were, and that light moves with a finite speed. When the same speed turned up in connection with electricity, Clerk Maxwell predicted electro-magnetic waves. Herz produced them, and Marconi put them at the service of mankind.

Modern astronomy, among other things, has given birth to spectroscopy. The spectroscope which analyses a beam of light into its component colours is the only means we have for investigating the composition of the stars, and it is largely for this reason that its use was developed. And it has turned out as practical an instrument as the telescope. It has been used in the analysis of minerals and the detection of poisons ; indeed, it has played its part in hanging several murderers. It is now throwing so much light on the structure of atoms and molecules that we may confidently hope that our grandchildren will learn a chemistry based on half a dozen simple laws instead of being compelled, like ourselves, to memorize the idiosyncrasies of the various elements and compounds.

But stellar spectroscopy has done much more than merely give the chemist a new method. It enables him to study matter under conditions of temperature and pressure which he cannot attain in the laboratory. If you want to know how a gas behaves at a pressure of a hundred thousandth of an atmosphere you can watch it in a vacuum tube in the laboratory ; if you desire to investigate it at a hundredth of that pressure, the astronomer will direct your telescope to a suitable nebula. And seeing that electric-light bulbs, X-ray tubes, the triode valves used in wireless, and the luminous tubes of sky signs all contain gas at low pressure, it is useless to describe the investigation of its properties as unpractical.

Astronomy began as the handmaid of astrology when men believed that the study of the heavenly bodies would enable them to predict events on earth. The old astrology is dead, but a few earthly phenomena have been found to depend on the sun and moon. To predict the height of the tides within an inch may seem an unnecessary refinement, but that inch may mean a saving of a hundredth of one per cent. in the expenses of a great port, and therefore be amply worth while.

And weather does to some slight extent depend upon sun-spots which appear according to a definite law. Attempts to predict the yields of crops by this method have met with small success, but the number of rabbits and hares in Northern Canada depends on that of sunspots to a remarkable degree. Every ten or eleven years the number of hares increases enormously, and a sudden pestilence then wipes them out. The next year there is great hunger among the lynxes and foxes which feed on them, and many more than usual are caught. It is quite safe to prophesy [1] that about 1926 there will be an abnormally large catch of red and cross foxes in Canada. And if the women voters can persuade the Government to appoint a national fur council, perhaps the price may come down.

[1] In 1925. I do not know if this prophecy has been fulfilled.

KANT AND SCIENTIFIC THOUGHT

IMMANUEL KANT was born at Könisgberg on April 22, 1724. He is one of the least readable of the great philosophers, and except in Germany is little read by scientific men who have at least a nodding acquaintance with a Berkeley, a Lotze, or a Bergson. But it is the purpose of this article to suggest that not only are his philosophical views of extreme importance for science, but that they are more important now than when Kant arrived at them a hundred and fifty years since.

The highest compliment which posterity can pay any thinker is to regard his most original thoughts as the data of common sense. In our time this has happened to Descartes. The average man would probably agree with him that matter had extension and mind none. He would use Descartes' brilliant invention of co-ordinate geometry to illustrate an argument on unemployment or climate. He would be willing to regard his body as a machine guided to some extent by an unextended mind. The main reasons for the triumph of Cartesian philosophy have been the apparent explanation of such properties of matter as heat, colour, sound, and odour in terms of its configuration and motion. The progress of physics until twenty years ago had thoroughly justified Descartes' apparently arbitrary interest in the spatial properties of matter. And similarly physiology seemed to be progressing

124

steadily towards an account of the body as a mechanism sometimes interfered with by a mind which could, however, for most purposes be left out of consideration. And whatever philosophical views one might subscribe to on religious or intellectual grounds, one tended to act over a large range of circumstances as if the above views were correct.

Except for Locke's distinction of primary and secondary qualities, very little post-Cartesian philosophy was incorporated into the assumptions of science, and the most recent work up to 1900, demanding, as it did, the postulation of an ether filling apparently empty space, bore a startlingly Cartesian appearance. In only one respect had any serious approach been made to the Kantian position. Mathematical physicists had quietly but definitely dropped the idea of causality; because they found that forces which have to be postulated as causes of motion do not possess those qualities of permanence which had rendered physical quantities such as mass, energy, and momentum so attractive. Of course, there were not wanting those who gave a more idealistic interpretation to the available evidence, but on the whole a realistic one seemed simplest. Then the theory of radiation broke down. It failed to explain radiation by very rapidly moving or very small bodies. The first failure led to the theory of relativity. According to this theory events form a four-dimensional manifold, and the relation between that series of events which constitutes our bodies and other series determines which of the latter we shall regard as simultaneous events, and which as successive and stationary. On Einstein's old theory the four-dimensional space-time was homogeneous, like the space and time of perception; and it was open to a philosopher who accepted his views to regard the action of the mind in perceiving space and time as merely selective, and not constitutive. But according to the general theory of relativity, which enabled Einstein to predict, among other things, the observed deflection of light by gravitation, space-time is not homogeneous, but bears a relation to the ' flat ' space-time of the special theory similar to that between the surface of an orange and a plane. If this is accepted (and scientific men in general accept it, because it enables them to predict certain observable phenomena with accuracy), it is clear that the action of the mind in perceiving homogeneous space and time is truly constitutive, and it is dubious how far the space-like character of the event-manifold is not a mere concession to our ideas of what a ' real ' world ought to be like. Eddington would go so far as to attribute every element in our experience of the external world, except that of atomicity, to our own mental processes, an interesting conclusion in view of Kant's insistence on the plurality of things in themselves.

The criticisms of the reality of space and time which arise from the theory of radiation by atoms are still more serious. The state of the atom before and after it radiates, and the subsequent history of its radiation, can be expressed in terms of the older physics, supplemented by relativity, with such accuracy that disagreements of less than one part in a thousand between theory and observation are the signal for a storm of further experiments. The probability of the passage of an atom from one stationary state to another, which coincides with the act of radiation or absorption, can also be dealt with by a mathematical theory due mainly to Planck and Bohr, and often with considerable accuracy. But every attempt to represent the process of radiation in terms of continuous space, time, or space-time, has broken down in the most hopeless manner. Bohr at least is convinced of the futility of any attempt at a ' model.' He is content to develop his beautiful, but highly formal, mathematical theory :—

' Und schreibt getrost ; Im Anfang war die Tat.'

And so the world of physics reduces to a manifold of transcendental events, which the mind distributes in space and time, but by so doing creates a phenomenal world which is ultimately self-contradictory. And this is approximately the position reached by Kant in the *Critique of Pure Reason.*

In biology we are for the moment in a curiously Kantian position. The mechanistic interpretation has nowhere broken down in detail. Every process in the living organism which has been studied by physical and chemical methods has been found to obey the laws of physics and chemistry, as must obviously be the case if, as Kant taught, these laws merely represent the forms of our perception and abstract understanding. But these processes are co-ordinated in a way characteristic of the living organism. Thus we cannot avoid speaking of the function of the heart, as well as its mechanism. Some biologists cherish the pious hope that the physico-chemical explanation will be found to break down at some point ; others the impious expectation that all apparently organic order will be reduced to physics and chemistry. There is very little in our present knowledge of biology to justify either of these standpoints, though evidence from other sources may seem to favour the former. The physiologist is therefore at present left in the peculiarly exasperating position reached by Kant in the second part of the *Critique of Judgment*. However mechanistic his standpoint, he must use the idea of adaptation at least as a heuristic principle. He will probably attempt to account for it as a result of natural selection, but natural selection is more fitted to explain the origin of given adaptations than the existence of living beings to which the term adaptation can be applied with a meaning. At present, with Kant, we are compelled to leave open the question ' whether in the unknown inner ground of nature the physical and teleological connection of the same things may not cohere in one principle ; we only say that our reason cannot so unite them.'

It thus appears that the doctrines of both physics and biology have reached stages which are more easily reconcilable with Kant's metaphysics than with that of any other philosopher. I do not suggest that either a physicist or a biologist need be a Kantian if he adopts any metaphysics : I claim, however, that other metaphysical systems, though they may be preferable on other grounds, are all definitely harder to adapt to the present data of science. If, for example, with Russell in his *Analysis of Mind*, we regard perception as essentially a selection of certain sensa from a larger number which exist, we arrive at a real world vastly more complicated than that of physics, even though it finds no room for purpose.   If, with J. S. Haldane, we regard purpose as more fundamental than mechanism, we have to look forward to a complete restatement of physics on teleological lines in the future, without being able to form any clear idea of how in detail this is possible.

I should be the last to suggest that the Kantian standpoint was any more final than the Cartesian.   On the other hand, there seems to me to be little ground for supposing that after another two centuries of scientific research (the conduct of politicians suggests that they may not be continuous) the data of science, which will then presumably include much of psychology, will support one rather than another of several post-Kantian systems.   And it looks as if Kant was at least correct when he claimed to have written the prolegomena to every future metaphysic.

The reason why Kant stands in this rather unique relation to scientific thought is probably that he was the last man to make contributions of fundamental importance both to natural science and to metaphysics. Apart from his work on meteorology and earthquakes, he was the first to put forward the nebular hypothesis, and to point out the importance of tidal friction in cosmogony.   He therefore understood the nature of scientific thought in a manner which is entirely impossible to the mere student of science and its history, and was able to frame a metaphysical system which is as applicable to modern scientific developments as the mathematical system of Gauss.   Until a first-rate scientific worker once more takes to philosophy we shall not see another Kant.

THOMAS HENRY HUXLEY

THOMAS HENRY HUXLEY was born at Ealing on May 4, 1825, but he was too great a man for his centenary to find him finally appraised and uncontroversially labelled. It was while in training for the medical profession that he published his first original work at the age of twenty. Entering the Navy as a surgeon, he was able, during four years in the tropics, to use his leisure to such effect as to revolutionize our views on the classification of the invertebrates. The importance of this work was instantly recognized, and at the age of twenty-five he was elected to the Royal Society. Although for the next twenty years he continued to work upon the anatomy of living and fossil animals, he had probably accomplished his best work in pure science by the age of thirty.

As lecturer, and later professor, at the School of Mines, in Jermyn Street from 1854 to 1872, and afterwards at South Kensington, he not only exercised an enormous personal influence on his pupils, but laid many of the foundations of the present methods of biological teaching throughout the world. His textbook of comparative anatomy is now, of course, out of date, but his *Human Physiology* is still perhaps the best book in the language for beginners in that subject. But it is on other grounds that he will be remembered outside scientific circles. In 1859 Darwin published the

130

Origin of Species, and Huxley, whom Lamarck and Spencer had failed to convince of the doctrine of evolution, was one of his earliest converts. He was far more of a fighter than Darwin. ' I will stop at no point,' he wrote, ' as long as clear reasoning will carry me further '; and whereas Darwin in the *Origin* was content to say, ' Much light will be thrown on the origin of man and his history,' and did not publish the *Descent of Man* until 1871, Huxley at once saw the implications of Darwinism in regard to the origin of humanity. It was at the Oxford meeting of the British Association in 1860 that he first entered the lists as a champion of man's animal origin against Bishop Wilberforce. The Bishop had concluded an attack on evolutionism by the question whether it was through his grandfather or his grandmother that Huxley claimed his descent through a monkey. ' If I am asked,' replied Huxley, ' whether I would choose to be descended from the poor animal of low intelligence and stooping gait, who grins and chatters as we pass, or from a man, endowed with great ability and a splendid position, who should use these gifts to discredit and crush humble seekers after truth, I hesitate which answer to make.' If this retort made little contribution to the solution of the problem of man's origin, it inaugurated a definite improvement in the manners of ecclesiastical dignitaries engaged in scientific controversy.

The next few years were largely devoted to a more serious defence of evolution, based on anatomical and palaeontological research, and summarized in *Man's Place in Nature*, published in 1863. The opposition

with which he met within the churches, and still more
perhaps the experience which he gained by service on
public bodies, ranging from the Royal Commission on
Fisheries to the London School Board, seem gradually
to have convinced him of the necessity of applying
scientific standards to every field of human activity,
including religion and education.   From 1870 onwards
he began to diverge from purely scientific themes into
fields of more general interest, and, like that of Voltaire,
his fame will rest largely on the production of his last
twenty years.

In particular, he conducted a sixteen years' controversy with Gladstone in the *Nineteenth Century* on
theological topics.   If the majority of educated Englishmen to-day reject the miraculous element in religion
and the infallibility of the Bible, the result is due to
Huxley more than to any other man, and in particular
to his extraordinary fairness of argument and moderation of language.   It is interesting to speculate on the
probable consequences had the protagonist in the fight
against religious dogma been a man of the type of
Bradlaugh.   In some directions the movement might
have made more progress.   Elementary education
might have been secularized, whereas Huxley supported
the teaching of the Bible in elementary schools.   One
of the great parties might have adopted an anti-clerical
programme ; but in such a case a large and compact
body, instead of an insignificant minority, would to-day
be supporting the religious ideas of 1860.

It was Huxley more than any one man who made
irreligion respectable in England.   To describe his
position he coined the word Agnosticism, denoting a

refusal to come to a decision on any question on which he considered the evidence to be inadequate. These questions included the existence of God and the immortality of the soul. ' If the condition of success,' he wrote to Charles Kingsley, ' in unravelling some little difficulty of anatomy or physiology is that I should rigorously refuse to put faith in that which does not rest on sufficient evidence, I cannot believe that the great mysteries of existence will be laid open to me on other terms.' This profound distrust of theories, however seductive, is characteristic of the experimental rather than the mathematical side of science, and Huxley's standpoint was very far indeed from the dogmatic atheism which often characterizes the mathematician who opposes religion. But if Huxley preserved an open mind on the metaphysical side of religion, he came to a definite decision with regard to its mythological aspect. ' To make things clear and get rid of cant and shows of all sorts. This was the lesson I learnt from Carlyle's books as a boy, and it has stuck to me all my life.' His polemical writing was largely directed against all allegations of breaches in the order of nature, from Noah's flood to spirit photographs.

He descended occasionally into politics, but his independence of judgment made it impossible for him to be a party man. He was one of a committee which urged the prosecution of Governor Eyre, of Jamaica, in 1866, for conduct somewhat resembling that of General Dyer in 1919, but in later life became a strong Unionist. While President of the Royal Society he refused to take part in politics, or even in such controversial movements as the Sunday League. His great gifts might have carried him as far in politics as Paul Bert was carried in France, but luckily for science he refused the offer of a seat in Parliament, and his last scientific paper was published only seven years before his death in 1895.

Perhaps his greatest defect as a thinker was his lack of sympathy with metaphysics. In this field he was a follower of Hume, Hamilton, and Mill, and never seems seriously to have considered the great movement which originated with Kant. That he was aware of the difficulties of his position is clear from his famous Romanes Lecture on ' Evolution and Ethics,' in which he contrasts the ethical process in man with the cosmic process of nature. He was a thorough believer in the absolute character of right and wrong, and far too honest not to see the difficulties in which this belief involved him. And his attacks on religion would have as little effect against a defence of it on metaphysical grounds as the metaphysical arguments of Dean Inge have on the average man. He fought his opponents with their own weapons, and proved that if religion is to be defended it is not on a basis of signs and wonders. Whatever additional facts may be true of humanity, it is subject to the same laws as those which govern the animals from which it has arisen. Until the mass of our people are convinced of this fact and ready to act upon it, Huxley's work will not be done.

WILLIAM BATESON

IF the Proceedings of the Brunn [1] Natural History Society had been a little rarer I suppose that Bateson would now be lying in Westminster Abbey. For we have only to read between the lines of the first report to the Evolution Committee of the Royal Society by himself and Miss E. R. Saunders, published in 1902, to realize that when Mendel's paper in the Brunn Society's journal was discovered in 1900, Bateson had already hit upon the atomic theory of heredity, which goes by the name of Mendelism. It was characteristic of him that no hint of this fact is to be found in his published work. His classical exposition of the subject is entitled *Mendel's Principles of Heredity*. Copernicus, if he admitted Aristarchus's priority, did not write on ' Aristarchus's principles of astronomy.' But Mendel's and Bateson's discovery was as fundamental as that of Copernicus, and of much greater practical importance.

And yet Bateson was not of a retiring disposition. The early days of Mendelism were marked by extremely violent controversy on both sides—I can remember the time when Mendelism was considered grossly heretical at Oxford—a controversy in which Bateson played a notable part. And his public attacks on the Darwinian theory were so phrased as inevitably to lead to the most heated argument, and even to the extraordinary misrepresentation that he disbelieved in evolution. So far

[1] Brno since the war.

was this from being the case that if he had died thirty years ago he would be remembered mainly for his work on Balanoglossus, a worm-like marine creature which he showed to constitute a link between vertebrates and invertebrates.

It was the wide and, as he felt, uncritical acceptance of the theory of evolution by natural selection which led him to expose its weak points. But it was eminently characteristic of him that he took up a not altogether dissimilar attitude to his own work. He had many disciples, but was never himself of their number. The characters which are inherited according to Mendel's laws are so numerous and important, and their possible combinations so enormous, that a lesser man would inevitably have devoted the remainder of his days to following out the detailed application of those laws. Bateson did so up to about 1912, but the last years of his life were largely given over to the investigation of exceptions to them ; and we owe to him more than to any other one man the demonstration not only that they are valid over a vast range of material, but that they occasionally break down. His last published work deals with these exceptions, and their importance is exaggerated rather than minimized.

His mental processes were well illustrated by his attitude to the work of Morgan and his school in New York, who have shown that the Mendelian factors are carried in or by the chromosomes which can be seen in a dividing nucleus. For eight years Bateson attacked this theory with the utmost vigour ; not because he considered it inherently improbable, but because he believed that it went beyond the evidence, and because the natural bent of his mind and his profound knowledge of the history of science led him to doubt the validity of long chains of reasoning, however convincing. When, however, the possibility of ocular demonstration arose, he went over to America, and returned a convert, though with certain reservations which I believe that the future will largely justify. It is the fact that he had retained his mental elasticity until the time of his death that makes that death so grievous a loss to biology.

Yet I can well believe that those who knew him but slightly carried away a different impression. He never attempted to conceal his contempt for second-rate work or second-rate thought, and pursued the truth with no more regard for other people's opinions than for his own. He started his career as a morphologist, and his outlook was always morphological. He was therefore sometimes unduly sceptical of reasoning from a non-morphological standpoint. But I never had an argument with him—and I had many—without the absolute conviction that he would no more hesitate to admit himself in the wrong if I could convince him, than to tell me that I was talking nonsense if, as was more usual, I failed to do so.

His scientific views inevitably led him to doubt the possibility of far-reaching improvements in human life by alteration of the environment. He was inclined to the belief that the best elements in the human race were being weeded out ; and the mutual destruction of them which went on during the war confirmed him in it. But he regarded most if not all of the attempts to apply science to this problem by creating an art of eugenics as premature in view of our profound ignorance of human heredity, and resolutely refused to associate himself with eugenical organizations. From the pessimism which such views inevitably engendered he found a refuge not only in science but in art ; and his exquisite sense of form drew him to the art of the far East, of which he was a well-known connoisseur.

If Bateson had merely demonstrated the truth and importance of Mendelian heredity the world would be his debtor. For in its essential manifestations it is so simple that I have known a child of fourteen apply it with complete success to practical breeding ; and yet it furnishes the only clue that we have at present to innumerable problems concerning the nature of the cell, the course of evolution, the determination of sex, and even the origin of certain human races.

But Bateson did much more than that. He has probably prevented Mendelism from becoming a dogma. For example, he held that it would not, as some at least of his disciples believe, explain evolution. It is normal for a discoverer to be obsessed by the importance of his own discoveries, and it is a thoroughly excusable weakness. There are times in the history of thought when an idea must be born, and if it is a great idea it may be expected to overwhelm and obsess the man who gave it birth. He either becomes its slave, or preserves a certain independence only by continuing to hold views incompatible with it at the expense of dividing his mind into watertight compartments. William Bateson escaped these fates because he was greater than any of his ideas.

THE FUTURE OF BIOLOGY

I N forecasting the future of scientific research there is one quite general law to be noted. The un-expected always happens. So one can be quite sure that the future will make any detailed predictions look rather silly. Yet an actual research worker can perhaps see a little further than the most intelligent onlooker. Even so, it may seem presumptuous for any one man, especially one who is almost completely ignorant of botany, to attempt to cover, however in-adequately, the whole field of biological investigation.

Every science begins with the observation of striking events like thunderstorms or fevers, and soon establishes rough connections between them and other events, such as hot weather or infection. The next stage is a stage of exact observation and measurement, and it is often very difficult to know what we should measure in order best to explain the events we are investigating. In the case of both thunderstorms and fever the clue came from measuring the lengths of mercury columns in glass tubes, but what prophet could have predicted this ? Then comes a stage of innumerable graphs and tables of figures, the despair of the student, the laugh-ing-stock of the man in the street. And out of this intellectual mess there suddenly crystallizes a new and easily grasped idea, the idea of a cyclone or an electron, a bacillus or an anti-toxin, and everybody wonders why it had not been thought of before.

139

At present much of biology is in the stage of measuring and waiting for the idea. One man is measuring the lengths of the feelers of 2000 beetles; another the amount of cholesterol in 100 samples of human blood; each in the hope, but not the certainty, that his series of numbers will lead him to some definite law. Another is designing a large and complicated apparatus to measure the electrical currents produced by a single nerve fibre when excited, and does not even look beyond the stage of the column of figures. If I were writing this article for biologists it would be largely a review of present and future methods; to a wider public I shall try to point out some of the results now emerging, and their possible application.

Let us begin with what used to be called natural history; the study of the behaviour of animals and plants in their wild or normal condition. Apart from animal psychology this has split up into two sciences, ecology and animal sociology. Extraordinary progress has recently been made in the latter. Wheeler of Harvard has made it very probable that the behaviour of social insects such as ants, instead of being based on a complicated series of special instincts, rests largely on an economic foundation not so very unlike our own. The ant that brings back a seed to the nest gets paid for it by a sweet juice secreted by those that stayed at home. Others, again, have been tackling the problem of how much one bee can tell another, and how it does it. To-morrow it looks as if we should be overhearing the conversation of bees, and the day after to-morrow joining in it. We may be able to tell our bees that there is a tin of treacle for them if they will fertilize those apple trees five minutes' fly to the south-east ; Mr. Johnson's tree over the wall can wait ! To do this we should presumably need a model bee to make the right movements, and perhaps the right noise and smell. It would probably not be a paying proposition, but there is no reason to regard it as an impossible one. Even now, if we take a piece of wasps' comb and hum the right note, the grubs put out their heads ; if we then stroke them with a very fine brush they will give us a drop of sweet liquid just as they do to their nurses. Why should we wait to see if there are ' men ' on Mars when we have on our own planet highly social and perhaps fairly intelligent beings with a means of communication ? Talking with bees will be a tough job, but easier than a voyage to another planet.

In ecology, where we deal with animal and plant communities which consist of many different species, each eaten by others from inside and outside, each living in amity with some of its neighbours, in competition with others, we are at present often lost in detail. But we are constantly finding that some hitherto unexpected but often easily modifiable factor, such as the acidity of the soil or the presence of some single parasite on an important species, will make a whole new fauna and flora appear, say an oak forest with wild pigs instead of a pine forest with ants.

We apply these principles in agriculture by using chemical manures and insects parasitic on those that attack our crops. But as we find the key chemical or key organism in a given association, we may be able vastly to increase the utility to man of forests, lakes, and even the sea. Besides this, however, one gets the very strong impression that from the quantitative study of animal and plant associations some laws of a very unsuspected and fundamental character are emerging ; laws of which much that we know of human history and economics only constitute special and rather complicated cases. When we can see human history and sociology against a background of such simpler phenomena, it is hard to doubt that we shall understand ourselves and one another more clearly.

In the domain of classificatory zoology our ideal is to establish a family tree of plants and animals : to be able to say definitely, let us say, that the latest common ancestor of both man and dog was a certain definite type of animal living, for example, in what is now the North Atlantic 51,400,000 years ago, under the shade of the latest common ancestor of the palm and beech trees, while the last common ancestor of the dog and bear lived only 5,200,000 years back. We are still thousands of years from this ideal, but we are now attacking the problem of relationships between living forms by a number of new methods, especially chemical methods. For example, we find that man agrees with the chimpanzee and other tailless apes, and differs from the tailed monkeys, in being unable to oxidize uric acid to allantoin in his tissues, as well as in many anatomical characters. This merely confirms our view that these apes are man's nearest relations. But the same kind of method will be applied to solving problems of relationship in which the anatomical evidence is less clear ; for example, what group of four-footed animals is most nearly related to the whale. Animals have a chemical as well as a physical anatomy, and it will have to be taken into account in their classification.

But the most important evidence about evolution is coming from the study of genetics. We take any animal or plant, and with sufficient time and money at our disposal should be able to answer the following questions (though if it is a slow-breeding animal like a cow it is more likely that our great-great-grandchildren will have to wait for the answer) :—

1. What inheritable variations or mutations arise in it and how are they inherited ?

2. Why do they arise ?

3. Do they show any sign of being mainly in any one direction, or of advantage to their possessor ?

4. Would natural selection acting on such, if any, as are advantageous, account for evolution at a reasonable speed, and for the kind of differences which are found between species (e.g., that which causes sterility in hybrids) ?

The first question can often be answered, the second rarely. Occasionally we can provoke mutations, as with radium or X-rays. There is no indubitable evidence that they ever arise in children in sympathy with bodily changes in their parents (the alleged transmission of acquired characters), and plenty of well-established cases where they do not. Now, we know how the genes, or units which determine heredity, are arranged in the nucleus of the cell, and also about how big they are. If we magnified a hen's egg to the size of the world (which would make atoms rather larger than eggs and electrons barely visible) we could still get a gene into a room and probably on to a small table.

But such magnification being impossible, the question how to alter a single gene without interfering with the others becomes serious, and some men have already spent their lives vainly on it ; many more will. The two most hopeful methods seem to be to find chemical substances which will attack one gene and not another ; and to focus ultra-violet rays on a fraction of a chromosome, the microscopic constituent of the nucleus in which the genes are packed. One can focus ultra-violet rays far more exactly than ordinary light, but even under the best conditions imaginable they would probably stimulate or destroy several hundreds of genes at a time.

Until we can force mutations in some such way as this we can only alter the hereditary composition of ourselves, plants, and animals by combining in one organism genes present in several, and so getting their combined effect. A great deal may thus be done with man. We know very little about human heredity as yet, though about hardly any subject are more confident assertions made by the half-educated ; and many of the deeds done in America in the name of eugenics are about as much justified by science as were the proceedings of the Inquisition by the gospels.

The first thing to do in the study of human heredity is to find characters which vary sharply so as to divide mankind definitely into classes. Most ordinary characters are no good for this purpose. We find every gradation of height, weight, hair, and skin colour. A few characters have been found, such as two which determine whether it is safe to transfuse blood from one man into another, which are definitely present or absent, and admit of no doubt. These are inherited in a very simple manner, and divide mankind into four classes.

Now, if we had about fifty such characters, instead of two, we could use them, by a method worked out on flies by Morgan of New York and his associates, as landmarks for the study of such characters as musical ability, obesity, and bad temper. When a baby arrived we should have a physical examination and a blood analysis done on him, and say something like this : ' He has got iso-agglutinin B and tyrosinase inhibitor J from his father, so it 's twenty to one that he will get the main gene that determined his father's mathematical powers ; but he 's got Q4 from his mother, to judge from the bit of hair you gave me (it wasn't really enough), so it looks as if her father's inability to keep away from alcohol would crop up in him again ; you must look out for that.'

When that day comes intelligent people will certainly consider their future spouses' hereditary make-up, and the possibility of bringing off a really brilliant combination in one of their future children, just as now we consider his or her health and education, before deciding on marriage. It is as certain that voluntary adoption of this kind of eugenics will come, as it is doubtful that the world will be converted into a human stud-farm.

The third question can be answered in the negative for certain forms at any rate. Out of over 400 mutations observed in one fly, all but two seemed to be disadvantageous ; and they showed no definite tendency in any one direction. But, of course, mutation may be

biased in other species.   The fourth question is largely
a matter of mathematics.   No competent biologist
doubts that evolution and natural selection are taking
place, but we do not yet know whether natural selection
alone, acting on chance variations, will account for the
whole of evolution.   If it will, we shall have made a
big step towards understanding the world ; if it will
no more account for all evolution than, for example,
gravitation will account for chemical affinity, as was
once believed, then biologists have a bigger job before
them than many of them think.   But a decision of this
question one way or the other will greatly affect our
whole philosophy and probably our religious outlook.

To turn now to the study of the single animal or plant,
physiological researches fall into several classes according to the methods used.   Some of us measure the
production of small amounts of heat or electrical energy
with complicated apparatus, others hunt down unknown
chemical substances, or measure accurately the amount
of already known ones in the tissues.   Taking the
biophysicists first, a whole new field has been opened
up by recent work on radiation.   When X-rays were
first applied to living tissues, it was very difficult to
get the same result twice running.   But now, thanks
to the work of our physical colleagues, we can get X-rays
of a definite wave-length and intensity, and our results
are correspondingly more intelligible.   In the same
animal one tissue is more sensitive than another to rays
of a given wave-length.   Moreover, cells are generally
more easily upset when engaged in division than at
other times.   These facts account for our occasional
success with X-rays against cancer, and our hope for

greater things in the future. It is quite possible that some combinations of invisible wave-lengths may be found to have special properties, just as a mixture of red and violet spectral lights gives us the sensation of purple, which intermediate wave-lengths do not.

Similarly, sunlight, besides warming us and enabling us to see, gives us bronzed skin, blisters us, wards off rickets, and cures many cases of tuberculosis. But are all these effects due to the same group of rays acting in the same way? We treat skin tuberculosis with ultra-violet light. Can we increase the curative effect without increasing the danger of severe sunburn? These questions are being answered as I write. The application of rays will gradually be taken out of the doctor's hands. He will write out a prescription, and we will go round to the radiologist's shop next door to the chemist's and ask for the prescribed treatment in his back-parlour. The next man at the counter will be after an apparatus to radiate the buds of his rose bushes during the winter, and kill off insect eggs which are out of reach of chemicals, without hurting the plants. The quack is already in the market with lamps producing radiation to cure rheumatism and make your hair grow. These are mostly harmless, though a few may be of value ; but probably the sale of X-ray tubes, which may cause cancer, will some day be as carefully regulated as that of strychnine.

Physical methods are also being applied in the study of the nervous system. We have by now gone most of the way in the localization of function there, for although a given area of the brain is always concerned in moving the hand, yet a given point in it may cause different

movements at different times ; just as any one tele-
phonist in an exchange can only ring up certain subscribers, but yet has a fairly wide choice. So we have
now got to work out the detail of the processes of
excitation and inhibition, as calling up and ringing off
are technically called. This involves very accurate
measurement of the electrical changes in nerve fibres
under different circumstances. Here we are still in
the graph and table stage, but probably only about ten
years off a fairly comprehensive theory of how the
different parts of the nervous system act on one
another. This will at once react on psychology, and
more slowly on normal life and practical medicine. A
great deal that passes as psychology is really rather bad
physiology dressed in long words, and the alleged
physiology in psychological text-books is their worst
part. We shall alter that. Until, however, we have
got a sounder neurology, scientific psychology, except
of a fragmentary character, is no more possible than
was physiology until chemistry and physics had reached
a certain point. And until psychology is a science,
scientific method cannot be applied in politics.

In chemical physiology we are after two rather
different things. The first is to trace out the chemical
processes in the cells, the nature, origin, and destiny of
each substance in them. The second, which is much
easier, is to trace the effect on cell life of various
chemical substances ; including those in which they
are normally found in the body, and unusual ones, such
as drugs and poisons. The first, if pushed to its logical
conclusion, would give us a synthetic cell, and later a
synthetic man, or ' robot.' The second would give

us a complete system of medicine, which is more immediately required. But, of course, the two react on one another and are not wholly separable.

At the moment the study of cell chemistry is leading to the most interesting results in the case of simple organisms such as yeasts and bacteria. For example, Neuberg of Berlin worked out a number of the steps in the transformation of sugar into alcohol and carbon dioxide by yeast; and was able, by appropriate chemical methods, to side-track the process so that it yielded other products. One of these is glycerine. During the war the Germans were unable to import the fats and oils from which glycerine is generally made. They needed glycerine for their propellant explosives, which contain nitro-glycerine. By getting yeast to make it from sugar they were able, in spite of the blockade, to produce all the nitro-glycerine they wanted.

This special process does not pay in peace-time, but there are others which do; and every day moulds and bacteria are playing a more important part in industrial chemistry. Similarly, we are now studying the chemical processes in bacteria as carefully as we do those in our own bodies. There is generally a weak link in such a chain ; for example, in human beings the links whose breaking gives us diabetes or rickets. If we study the tubercle bacillus carefully we may find his weak point. The relatively direct methods which gave us the cure for syphilis are here no use, for the tubercle bacillus is a far tougher organism than the spirochaete, and we cannot yet kill him without killing his host. Similarly, we are trying to find out how the chemical processes in normal and cancerous cells differ.

In man the study of what our body cells can and cannot do is gradually leading us to the perfect diet. It is becoming quite clear that faulty diet gives us some diseases, including most of our bad teeth, and predisposes us to others ; and that nothing out of a tin or package so far comes up to natural foodstuffs. On the other hand, as the population of large cities cannot get these, we have got to determine what can be done to improve a diet based largely on milled cereals and tinned milk and meat. It is a tough problem, and for every pound we can spend on research and publicity put together the food-faking firms have a thousand for advertising of ' scientific ' foods.

To turn now to the chemical co-ordination of the body, we know that various organs secrete into the blood substances (often called hormones) which profoundly affect the rest of the tissues. A number of these have been obtained in a fairly concentrated form —that is to say, mixed with perhaps only ten or a hundred times their weight of other substances. Only two have been obtained entirely pure, though presumably all will be. Now, if we take one of the most widely popularized of recent therapeutic methods, the grafting of apes' testicles into old or prematurely senile men, this is just an attempt to get a hitherto unisolated hormone into the blood stream. The operation is expensive, the idea unpleasant, and the graft generally dies in a few years at most. The problem is to isolate the hormone free from other poisonous substances found in most tissue extracts, and later to find its chemical formula and synthesize it. One of the corresponding substances found in the female sex has been obtained free from harmful companions by Allen and Doisy in America.

When we have these substances available in the pure state we ought to be able to deal with many departures from the normal sexual life, ranging from gross perversion to a woman's inability to suckle her children; since lactation, as well as the normal instincts, appears to depend on the presence of definite substances in the blood. We shall also probably be able, if we desire, to stave off the sudden ending of woman's sexual life between the ages of forty and fifty. It is worth pointing out that there is no serious reason to believe that any of the rather expensive products of the sex glands now on the market, and often prescribed by doctors, are of any value except as faith cures.

A much more ambitious attempt to deal with old age is being started by Carrel. Cultures from individual cells from a chicken can be kept alive in suitable media for twenty years, and as far as we know for ever. To live they must have certain extracts of chicken embryo. The blood of a young fowl contains substances (which can be separated by suitable methods) that both stimulate and check their growth. The former is absent in very old fowls. The problem of perpetual youth has, therefore, been solved for one kind of cell. But to make a pullet immortal we should have to solve it for all the different cells of its body at once. We do not know if this is possible, or whether it is like trying to design a society which is ideal alike for cowboys, automobile manufacturers, and symbolist poets, all of whom can hardly flourish side by side. Fifty years hence we shall probably know whether it

is worth seriously trying to obtain perpetual youth for
man by this method. A hundred years hence our
great-grandchildren may be seeing the first results of
such attempts.

Besides these rather sensational substances which
were first detected by their effects on organs, the proper
working of the organism depends on the amount of
quite well-known bodies, such as sugar, oxygen, and
lime in the blood. We are gradually getting to know
the amounts of these required for health, but it is much
harder to estimate the amount needed of such a substance as, say, insulin. We can now kill an animal
and produce a fluid from inorganic constituents that
will keep its heart or liver alive for a day or more.
Soon it will be a matter of months or years. To keep
tissues alive for a time comparable with the life of their
owner we shall have to have about 100 substances, but
perhaps not very many more, present in the normal
amounts in the fluid perfusing them. At present we
only know the correct quantity of some twenty, if that.
Given this knowledge and the means of applying it, we
could make good the deficiency of any organ but the
nervous system. We could grow human embryos in
such a solution, for their connection with their mother
seems to be purely chemical. We could cut our beefsteak from a tissue culture of muscle with no nervous
system to make it waste food in doing work, and a
supply of hormones to make it grow as fast as that of
an embryo calf.

In pharmacology our knowledge rests mainly on a
series of lucky accidents. A few of the complicated
substances made by plants have a striking effect on

animals, but why a molecule of a given build has a given physiological effect we are only beginning to discover. When we know, we should be able to make as great an advance on plant products as we did with dyes when the relations between colour and chemical composition were discovered. If we had a drug that was as good a pain-killer as morphine, but one-tenth as poisonous and not a habit former, we could use it indiscriminately ; and wipe out a good half of the physical pain in human life at one stroke.

Such are a few of the possibilities of our science. It is easy enough to say what we would do if we had a method to measure A or isolate B. But it is in inventing and applying these methods that our biggest problems often arise.

NATIONALITY AND RESEARCH

SCIENCE is an international concern. Any paper on pure science becomes the property of the whole world the moment it is published. And the special scientific terminology so frequently termed jargon is, with all its faults, an international language. One can get the gist of a scientific paper in any European tongue, and even amid a wilderness of Japanese script one comes across oases of mathematical expressions, numerical tables, and chemical formulae. Moreover, all important papers are abstracted in English and German within a year or so of publication.

It is impossible for any one critic to assess the contributions of the various nations to literature. For all I know, the greatest living poet may write in Siamese. Musical notation is more nearly international, but it must be remembered that many Oriental peoples employ a scale very different from our own. In the domain of science one may safely be more positive, and an attempt to apportion the contributions made by different nations, if unlikely to be wholly impartial, is not obviously futile.

As elsewhere, one immediately comes up against the problem of the Jews. Are we to call Einstein, who is a professor in Berlin (and also in the Dutch University of Leyden) but was born in Switzerland, and is international in outlook, a Jew, a German, or a Swiss ? For the Jews, just as they are partly responsible for one of the worst features of our civilization, the control of industry by financiers more interested in profit than service, have shown in other fields the most single-minded devotion to pure thought. ' German ' science in the last forty years has been largely Jewish, in spite of the very unfavourable conditions under which the Jews worked. Thus Ehrlich's co-discoverer of ' 606 ' was a Japanese, Hata, because so few German gentiles were willing to work with him. Mendeleeff and Metchnikoff, the two greatest Russian scientists of last century, both had Jewish mothers. So far the main Jewish contribution to science has been in Germany and Austria, but it is beginning seriously in Britain, America, France, and even Japan, while the first papers from the biochemical laboratory of Jerusalem University were published in 1925.

As modern science is of European origin, it will perhaps be convenient to work in towards Europe as a centre from the rest of the world. New Zealand and Australia have made first-rate contributions to science, but largely by exporting their scientists to other portions of the British Empire. Rutherford, who discovered the structure of the atom, was born in a back block of New Zealand, and gravitated to Cambridge via Montreal and Manchester.

Japan is making contributions to every branch of science, but, as a student of Japanese art might expect, they have, on the whole, been distinguished by technical rather than intellectual power. For example, in bacteriology Japan holds a very high place ; in physics it excels rather in exact measurements than in their theoretical interpretation. However, Japanese research work is still in its first generation, and is already ahead of that of most European countries. In another fifty years it may excel that of Europe as a whole.

China is starting on research, largely under American guidance. India has begun, and that sensationally enough. Srinavasa Ramanujan was a clerk in a Madras office with no mathematical education beyond that usual in secondary schools. In 1913 he sent to Cambridge proofs of certain new theorems in higher algebra. He was at once brought over to England, and within a few years he was a Fellow of the Royal Society and of Trinity College, Cambridge. Had he lived a century earlier, when the methods which he favoured were yielding their best results, he would probably have been the world's greatest mathematician. And though he died too early to earn that title, he may perhaps be awarded the palm for mathematical originality in the twentieth century. India has produced no other scientist of such distinction, and her total contribution has been less than that of Japan. But it has been, on the whole, of a surprisingly original character, sometimes, indeed, slightly bizarre to European minds, and leaves no doubt whatever that India has a very great scientific future.

South America has as yet done little, though the Argentine Republic has not been without its distinguished biologists and palaeontologists.

The United States produce a colossal volume of scientific work, of very unequal merit. Where endowment can assure results, they lead the world. Their astronomical observations form the bulk of international output, though their interpretation often comes from England, Germany, or Holland. In the studies of animal breeding and nutrition, the methods largely devised in Cambridge and London are being developed on a colossal scale. Morgan's work on inheritance in New York has involved the counting of over twenty million small flies. Langmuir is provided by the General Electrical Company with his own laboratory and a salary which most Cabinet Ministers would envy, on condition that he occasionally spares a day or so to consider the problems which arise in their works. Some of the ablest men in Europe are constantly being attracted over by offers of salaries, and still more of facilities of research.

In spite of these facts and the undoubted genius of many Americans, I am inclined to think that in pure (though not perhaps in applied) science America produces less than either Britain or Germany. The probable reason is that great men are more important to science than great laboratories, and a larger proportion of scientifically-minded men are drawn into the work of national development in America than in Europe. The very wide diffusion of higher education in the U.S.A. is compensated for by its often indifferent quality, and by the terrific obscurantism which makes biological teaching a farce in many parts of the country.

Though Canada has sent fewer notable men of science to Britain than Australia, her output of published work, culminating in the preparation of the internal secretions of the pancreas and parathyroid by Banting and Collip, has been greater, partly owing to a constant interchange of ideas and personnel with the United States.

The Union of Socialist Soviet Republics (I do not say Russia, if only because of the admirable physiological work proceeding in the Georgian University of Tiflis) is still so isolated that its appraisal is difficult. During the war and revolutions about one-third of its scientific personnel appear to have been lost by death, flight, and the separation of Poland and other states. I can only think of one biologist of any eminence executed by the present regime, and could cap his name with that of another recently killed by Whites in the Caucasus. As a result of the revolution, scientific workers have been given many large houses as laboratories and museums. On the other hand, those in the higher positions are worse off economically, and all of them are largely cut off from foreign sources of literature and equipment. Hence they are concentrating on work where elaborate apparatus is not required, for example animal breeding, to which they appear to be devoting more effort than any country but the United States, and the careful study of animals and plants in a state of nature.

In ' bourgeois ' Europe two of the small nations, Holland and Denmark, undoubtedly lead in the output of scientific work per million inhabitants, though Switzerland runs them very close. They are incidentally two of the healthiest nations in Europe, although the population of Holland is increasing faster than that of any other European state ; and are both quite rich, although almost devoid of mineral wealth. This is because they are successfully applying biology— Denmark to her own agriculture, Holland to the development of her empire, which sets the example to the world in tropical agriculture and hygiene. The Dutch go so far in their recognition of this fact as actually to call some of their liners after botanists, instead of the usual soldiers, sailors, politicians, and royalties.

In Denmark the Carlsberg brewery, which, under the supervision of two great biochemists, has come to produce the strongest beer on earth, is now the principal support of a scientific academy, as though ' Bass ' were affiliated to the Royal Society !

The largest actual output of scientific work comes from Britain and Germany. Fifty years ago, Germany probably took the first place in this respect. To-day I am personally inclined to think that the position is about equalized, and that this was so even before the war. The reasons are perhaps as follows. Before the foundation of the German Empire in 1871, each little state had its university, which competed with its neighbours and managed to make any of its particularly brilliant alumni a professor extraordinarius at an early age. He then gathered pupils round him and formed a school. Now he probably goes to the Kaiser Wilhelm Institut at Berlin and has to compete for pupils with colleagues from all over Germany. Moreover, there is reason to think that politics had come to play more part in university appointments in the last few decades before the war than formerly. Nevertheless, it would be idle to deny the splendour of Germany's achievements at the present moment, more especially in such fields as organic chemistry and mathematics, which the Germans have made peculiarly their own.

If Great Britain leads the world in many branches of

science, it is, I think, largely through two causes, the autonomy of our universities, and the lack of nationalism in our scientific thought. A university governing itself may be a little deaf to the claims of working-class education or the equality of the sexes, but it is more likely to appoint the best man to a post than is one governed by business men or politicians. And we are certainly less prone than France, Germany, or America to ignore the work of foreigners.

France has undoubtedly lost ground in the last fifty years. She still produces men of very great originality, but State control of higher education and ignorance of foreign achievement handicap them enormously. Since the war, moreover, no serious attempt has been made either in France or Belgium to bring the real wages of scientific workers to pre-war levels, and many of their best men are turning to applied sciences such as medicine and engineering.

Sweden, Norway, Switzerland, Belgium, and Finland are all producing first-rate work, and the same applies to Poland, Czecho-Slovakia, Austria, and Hungary, which have shared most of the scientists of the late Austrian Empire. Italy is now producing little experimental science, though her mathematics are still very good indeed, the *Rendiconti del Circolo Matematico di Palermo* being one of the world's greatest mathematical journals. Though Spain has given us a great microscopist and Greece a great mathematician, our story is told, and it only remains to see if we can draw any lesson from the distribution of scientific achievement.

Other things being equal, the small nations are more scientific than the large, for reasons already discussed.

Probably a standard educational system is an evil, as government officials always tend to demand quantity rather than quality of work, and research flourishes best in an atmosphere where leisure and even laziness are possible. On the other hand, a government department like the Medical Research Council in England, which is not dominated by red tape and is willing to subsidize work that may turn out to be valueless, on the chance of obtaining a really great result like the recent production in a pure form of the vitamin which prevents rickets, can be of enormous use to science.

Scientific ability is not the perquisite of any one race but it can only show itself under conditions when thought is free, and there are many different ways of suppressing it. One way is to refuse research facilities to people without academic qualifications. If Faraday lived to-day he would not find his career much, if at all, easier in England ; and in many countries he would have to remain a bookbinder. Scientific genius is so rare that no single system, however well thought out, will avail for its discovery and encouragement

SCIENTIFIC RESEARCH FOR AMATEURS

U NTIL the last century scientific research was almost entirely the work of men who earned their living by some other method, or possessed private means. Until about fifty years ago there was no such thing as training for research, and every researcher began his career as an amateur. To-day large sums of public and private money are devoted to the education of prospective research workers. This expenditure is most valuable in certain branches of science, but as one result of it the view has unfortunately got abroad that research is now only possible to persons who have gone through an elaborate and expensive training. The same false idea is propagated by writers who are ignorant of science and either hate or fear it, and therefore attempt to magnify the gulf which exists between the scientific worker and the average intelligent man. It is the object of this essay to show that any man possessed of the patience and leisure necessary to watch a cricket or baseball match, and sufficient intelligence to solve a crossword puzzle, can make quite definite contributions to scientific research.

It must be admitted at once that certain branches of science are almost closed to the amateur. In chemistry, for example, or human anatomy, the would-be researcher must not only master a great deal of knowledge, but, what is far more serious, and indeed almost impossible outside the laboratory, a great deal of technique.

The reason for this is fairly simple. There are only ninety known kinds of atom, and only two main types of human body. Each (except half a dozen excessively rare types of atom) has been very extensively studied. But several hundred thousand different species of insect and several million different stars are known ; and very little indeed is known in detail about most of them, while all would repay study.

It is probably in the biological field that the amateur can do the most interesting and valuable work. If he lives in a town his possibilities are restricted to a study of those organisms which can live under artificial conditions. But the number of such is quite large. Enormous numbers of town dwellers keep animal or plant pets, and many breed them. Now exceedingly little is known as yet with regard to inheritance in most plants and animals. The animal breeder would perhaps be well advised to avoid such relatively thoroughly studied species as mice, rats, rabbits, guinea pigs, and poultry ; but a vast amount remains to be done with dogs, cats, and pigeons. With regard to dogs, the inheritance of colour within some breeds is fairly well understood, but singularly little has been found out as yet concerning the structural characters which distinguish the breeds from one another. Some can be transferred as units, so to speak. A Swiss breeder produced a Newfoundland with the legs of a Dachshund ! Others are more complex. The same is true of pigeons. So far the ambition of pigeon breeders has more often been to become a prominent member of the Bald and Beard Club or the Oriental Frill Club than of the Genetical Society. But the production of the ideal

Bald and Beard pigeon does not constitute a permanent acquisition to humanity, as would the knowledge of how its baldness and beard are inherited. And as a matter of fact it is often much easier to determine the inheritance of a character than to establish a strain of prize-winners. Science is indebted to the fancier for picking out and perpetuating a number of remarkable varieties. It is now beginning to turn to him to analyse his truly wonderful material.

Moreover, there are many animals as easy to breed as pigeons and cats, and far less troublesome to keep, including many insects and molluscs. One of the Clerks to the House of Commons has devoted his spare time for the last few years to the breeding of water snails, which requires no apparatus more complicated than jam pots, water and water weeds. Some of these snails coil in the wrong direction, *i.e.* look like the mirror image of an ordinary snail ; and this character was found to be inherited in a wholly unprecedented manner. To establish the laws governing this new type of inheritance over 50,000 snails had to be bred, and the work was almost entirely done by a few amateurs.

Any one with a garden or greenhouse, however small, can embark on plant breeding. Apart from plants of economic value only a very few, such as the sweet pea, stock, snapdragon, and *Primula sinensis*, have been thoroughly studied from the point of view of heredity. The technical equipment needed consists of a brush to carry pollen from one flower to another, a pair of fine scissors to remove stamens, and some paper bags to keep out unwanted pollen.

It might be thought that the results of such researches could only be of trivial importance. This is not so. Even if one does not hit upon facts which involve a new type of inheritance or throw new light upon one already studied, the facts discovered have a comparative value. As we study the variations in different species of plants and animals and the laws governing their inheritance we begin to get a view of the raw material upon which natural selection has worked to produce evolution.

The more ambitious student of biology, especially if he or she combines scepticism and pertinacity, can study inheritance in human beings. Here experiment is impossible, and facts often hard to obtain. Naturally enough people will lie with regard to the inheritance of a mental, moral, or physical defect. But experience shows that they lie almost equally with regard to such a harmless topic as the eye colour of their late grandmother; apparently from a desire to give some kind of an answer to a question which they are convinced is quite unimportant. First-hand evidence is therefore always desirable and often necessary. And so mobile has the human race become in the last generation that the tracing of a single family may demand railway journeys running into thousands of miles. It is cheaper to breed armadillos or cacti, or even, like Professor ———, to trust largely to one's imagination so far as human inheritance is concerned.

In the country the amateur biologist can observe as well as experiment. The natural history of our grandparents' time, which was regarded as unscientific by our parents, is now coming into fashion again. The collector is very unlikely to discover a new species, except in undeveloped countries, but he may contribute to our knowledge of the geographical distribution of an old one, or still more effectively to an understanding of the conditions under which it flourishes. Ecology is the study of natural communities of plants and animals. The ecologist studies not only the conditions of soil and climate which determine that one community rather than another should exist, but the relations between the different members of the community ; the most important and typical relation being that of eater and eaten. This is a harder job than it sounds. Even in your own garden you may know the principal caterpillar that eats your favourite flower, and the birds that eat the caterpillar and thus protect the flower ; but probably far more caterpillars are eaten from inside by parasitic insects than from outside by birds. So a complete ecological study of any area must be quantitative, and is as different from the old-style natural history as an economic survey from a list of trades and professions. Nevertheless, any lover of country life who keeps his eyes open can make some contribution to the science.

One of the most universal and remarkable of human characteristics is a love of flowers. Nothing can be more delightful than to pass a quarter of an hour daily wandering round one's garden or favourite fields and hedgerows, noting what new flowers are to be found in bloom. A diary in which records of blooming are kept year after year is not only an ideal hobby, but of considerable scientific value. Such diaries are collated in England by the Phenological Society, and furnish one of the most important means available for the study of the influence of climate on plant life. The variations in flowering time from place to place and from year to year, combined with records of temperature, rainfall, and sunshine, show how the various types of plant respond to their environment. But even in England far too few records of this kind are kept, and elsewhere the need for them is much greater.

The naturalist who can spare more time together may embark upon the study of animal behaviour, and more especially insect behaviour. I do not use the word psychology because it is not fashionable, although to my mind quite justifiable. Here insects and spiders with their very definite and complicated behaviour patterns, or ' instincts,' are the ideal objects of study. A lens, a notebook, and immense patience, are the chief requisites, and it is worth noticing that many of the greatest observers of insect behaviour had other occupations. Fabre was a schoolmaster, Lord Avebury an extremely successful banker and a politician ; in fact, until quite recently professional biologists had contributed relatively little to this branch of knowledge, and some of them had blundered pretty badly on leaving the laboratory for the field.

The other branches of biology, although open to the amateur, are mostly less promising because they require more technique ; and the same is true of chemistry and physics. Here the amateur is more likely to invent than to discover. Accurate measurements are almost always needed in work that is to be of theoretical value, whereas a combination of luck and ingenuity may easily lead to an improvement in some well-known process which is of great practical importance.

In geology the amateur can do a great deal in relatively undeveloped countries, rather less in those which have been thoroughly explored ; but one branch of it, namely, human palaeontology, has been traditionally the field of enthusiasts from other walks of life, especially of Roman Catholic clergy. Local knowledge has, of course, been of the greatest importance, and the exploration of the underground haunts of primitive man has called not only for skill and perseverance, but in many cases for considerable courage.

Among all the achievements of scientific amateurs none is more epic than the swim of Norbert Casteret through the cavern of Montespan. Casteret, who is one of France's champion swimmers, conceived the idea of swimming up a stream which emerges from a cave in a part of south-western France where several primitive human races have left their bones and tools. Carrying candles and matches in a waterproof bag he went up the cave until the roof met the water, dived, and swam on, holding his breath, until he found air once more above him. He was now in a section of the cave which had been sealed off from the world by water since palaeolithic times. Walking or swimming along it and diving under the rock where necessary, he finally emerged into the daylight more than a mile from his starting point. The results obtained were of the first importance, for he discovered several rude palaeolithic statues of unbaked clay, of a type entirely new to science. As professional students of primitive man rarely, if ever, combine phenomenal powers of diving with superb courage, this particular type of research is likely to remain the prerogative of the amateur

Spelaeology, as the science of caves is called, is a

sport as fascinating and arduous as mountain climbing, and in Europe and Asia at any rate leads its devotees into the homes of earlier human races. One may hope that as virgin peaks decrease in number, sportsmen with a scientific bent may begin to feel the lure of the virgin cavern. Unfortunately those English caves which are most interesting to the pure spelaeologist, rarely contain human remains. One may hope that the descendants of the British sportsmen who conquered so many of the Alpine peaks will turn their attention to the limestone caverns of Yugoslavia and north-eastern Italy.

A great deal of meteorological observation must inevitably be made away from great centres of population. Even in England the distribution of rainfall is far less known than it should be, and in less populated countries accurate knowledge is ludicrously inadequate. The reason perhaps lies in the fact that only continuous and methodical observation is of the slightest value. The rain gauge must be read day after day at exactly the same hour, and the wetter the day, the more urgent the need of punctual observation ! Yet of all forms of amateur scientific work the daily reading of a rain gauge bears the greatest resemblance to the daily routine of the laboratory, where many thousands of accurate observations may be required to establish a fact whose significance is even then doubtful.

Observations on wind, sunshine, and clouds are, of course, also needed ; but whereas a daily visit to the rain gauge is generally sufficient, a day's anemometer records must be carefully kept, and the clouds change from minute to minute. Nevertheless, the lover of photography may do worse than turn his attention to them; either using a special cloud camera which gives a picture of the whole sky on one plate, or a pair of cameras some hundreds of yards apart, and giving stereoscopic pictures.

But it is in the region of the atmosphere somewhat above the clouds that the amateur observer reigns supreme. ' Shooting stars,' which are due to fragments of matter from interstellar space flaring up through friction as they enter our atmosphere at very high speeds, have been known from the remotest antiquity; but we are no better able to observe them to-day than were the wise men of the East, or Mohammed, who thought that they were missiles cast by angels at devils who were attempting to overhear the secrets of heaven. The telescope is far better suited than the naked eye for the observation of ordinary stars, but it greatly magnifies a very small portion of the whole heavens, and hence the chance of a shooting star crossing its field is small.

Some day, no doubt, an instrument superior to the human eye will be invented for the observation of meteors; but until this invention is made, the equipment of an observatory will be worse than useless for this purpose. Hence the world's greatest meteor observer, and probably its greatest amateur scientist, Mr. W. F. Denning, owns no allegiance to any observatory. The observer of meteors requires a clear sky, a thick coat, a notebook, knowledge of the constellations, infinite patience, and a tendency to insomnia. Our knowledge of the upper atmosphere has already been revolutionized by mathematical physicists using Mr. Denning's observations made over a period of more than thirty years. They show that at a height of forty miles the air is many times denser than was formerly supposed, and attains the quite comfortable temperature of 85° F. If this were not so, meteors would mostly come far nearer to the earth's surface, and would therefore be far brighter than is actually the case. However, in order to give a really precise basis for calculations vastly more observations of the same kind are required. The hot layer is probably the same as the ionized Heaviside layer which reflects downwards the Hertzian waves used in radio-telegraphy and telephony. The heating and ionization are caused by the stoppage of some of the ultra-violet components of the sun's radiation.

The astronomer who is able to spend the price of a small motor car on a telescope can at once enter on work of a character demanding less patience. A specially suitable field is the observation of variable stars. Many stars vary in brightness, and the variation is generally periodic. A knowledge of the way in which their luminosity varies with the time throws much light on their nature. The variation may be due to a periodic eclipse by a darker star rotating round the star observed and occasionally cutting off its light, or to real changes in the luminous star itself. Many variable stars appear to pulsate like a beating heart with a period of hours or days. The fact that this period is increasing (though very slowly) in some variable stars, is the only direct indication we have of real changes in the nature of the fixed stars ; and supports the view held on many theoretical grounds, that, to use a perhaps rather inaccurate metaphor, they are running down. Amateur astronomy is well organized both in England and America, and boasts of the only paper that sets cross-word puzzles in three dimensions. There are so many stars that a useful field of observation could be found, not perhaps for the whole population of the world, but certainly for some hundreds of times the present number of astronomical amateurs.

Mathematics is the most abstract and specialized of the sciences, and in view of the very high qualifications required in a professional mathematician, might seem a hopeless field for all but a very few. And certainly the amateur cannot hope to rival the professional at his own game. Nevertheless, there is an opening in mathematics for the class of mind that delights in numerical calculation for its own sake. Those who are so gifted tend to amuse themselves with calculations which are useless ; for example, of the number of grains of sand of a given size needed to fill the dome of St. Paul's ; or worse than useless, as when the date of the outbreak of the next world war (May 1928) is deduced from the dimensions of the great pyramid. If more mathematically minded they might occupy themselves somewhat unprofitably in calculating the ratio of the circumference of a circle to its diameter to a thousand places of decimals instead of a mere seven hundred odd, its present limit of accuracy.

But they can also be of real if limited value to mathematics in two ways. They can work out tables of certain mathematical functions which are not as yet tabulated, but which are sometimes needed by physicists, engineers, and others, in their calculations.

And they can study the properties of whole numbers. In mathematics as a whole theory runs ahead of observation. But in the theory of numbers this is not so. No one has ever been able to prove, for example, that every even number greater than two can be expressed as the sum of two primes. Yet this is as well established by observation as any of the laws of physics. It is known that this and various other theorems are true if a certain hypothesis about the Zeta function, enunciated by Riemann nearly a century ago, is correct. No one has been able to prove this hypothesis. It has only been shown that all the consequences deducible if it is true are so far verified by experience. But any day a computer with little knowledge of pure mathematics may disprove it. Here then is a possible triumph for the mathematical amateur. Similarly while it has been proved that the circle cannot be squared, the possibility remains open that the sum of two fifty-ninth powers may itself turn out to be a fifty-ninth power ; just as, for example, the sum of three squared and four squared is itself the square of five ; although since the seventeenth century the opposite has appeared likely, and has been proved for all powers less than fifty-nine. Only a few years ago Mr. Powers, an American computer, disproved a hypothesis about prime numbers which had held the field for more than 250 years. Here then is a relatively hopeful field for circle squarers.

And it is to amateurs that the world must look for perhaps the greatest of all scientific achievements, the foundation of a new science. The founder may be an amateur in the strictest sense, as was the abbot Mendel when he founded genetics, or a worker at some other science, like the physiologist Galvani, who discovered electric currents ; but he is always an amateur in the science that he creates.

And just because in the modern organization of research any given worker is labelled as an astrophysicist, an organic chemist, a palaeontologist, a geneticist, or what not, the need for the amateur in the border line or wholly novel subject is all the greater. Fortunately, most of the very greatest scientists keep the spirit of the intelligent schoolboy to the last. Darwin occasionally indulged in what he called ' fool's experiments,' as when he played a trumpet to a group of climbing plants.   In this case nothing happened, but the experiment was no more intrinsically foolish than that of Oersted when he tried the effect of an electric current on the compass, and thus connected electricity and magnetism.

The objection to most amateur science lies not in the foolishness of its experiments, but in the inability of the experimenters to be satisfied with negative results. Most laboratory experiments are failures, and even when an apparent success has been obtained the competent researcher at once tries to catch himself out.   I am going to waste to-morrow on an experiment which I hope and trust will be a failure, for if it were a success it would not only be quite inexplicable, but would destroy the theoretical results of a year's work. Amateur scientists commonly fail because they set out to prove something rather than to arrive at the truth, whatever it may be.   They do not realize that a good half of most research work consists in an attempt to prove yourself wrong.   Intellectual honesty is discouraged by politics, religion, and even courtesy. It is the hardest but the most essential of the habits which the scientist, whether professional or amateur, must form. And if he can spread the habit among his fellow-men it may prove to be a contribution to the good life compared to which the applications of science to engineering and medicine are comparatively unimportant.

SHOULD SCIENTIFIC RESEARCH BE REWARDED?

IT is a commonplace that the wealth of modern societies is in a large degree the creation of scientific research, and it is therefore often suggested that scientific workers should share in this wealth on a scale more or less comparable with that in which labour and capital are rewarded. In particular, Sir Ronald Ross has for some years supported the idea that scientific workers who have made important discoveries should, as a routine, receive pecuniary rewards from the State—of course, the State—on a scale comparable with those of successful generals.

At first sight the justice of his contention seems clear. In other intellectual occupations—for example, literature, music, and the fine arts—those men and women who are most highly esteemed by their contemporaries enjoy fairly large incomes. A scientist can practically never hope to earn as much as £2000 per year as such, and is generally glad if he gets half that sum, while incomes of £300 are quite common. To make more he must go out of the path of pure research and teaching, and apply his talents to industry, administration, popular writing, or some other activity for which there is an economic demand. Very little can be made by taking out patents. One cannot patent a new fish or a new star, and medical etiquette rightly forbids the patenting of a new medicine or a new surgical instrument. The physicist and chemist cannot patent their

176

great discoveries, for two different reasons. In the first place they do not know how they will be applied. When Richardson discovered the laws governing electron emission from hot metals, he did not, presumably, suspect that he had made wireless telephony practicable. He could not take out a patent for all possible applications of his discovery ; and many equally fundamental facts elicited at about the same time—for example, the emission of electrons by radio-active bodies in the cold —have found no paying application as yet.

Again, as a general rule, the greater the discovery the longer the time before it is applied. It was a long time before Faraday's discovery of electro-magnetic induction could be applied to the manufacture of a practical dynamo. He did not make a penny out of it, and, fortunately, his heirs-at-law are not still being paid for it, as are those of Nelson for the Battle of Trafalgar. The Chinese might regard it as equitable that every user of electric light or power should burn at least one currency note per year before Faraday's image ; and even a posthumous reward is better than none.

The French, for example, do not pay their men of science a living wage. On the other hand, they give them statues—often very good statues—when dead, and call streets after them. And to many scientific men, such is our perversity, the prospect of becoming a street name is a better incentive to effort than a rise of salary.

The greatest difficulty of a scheme of rewards during life is to be found in the impossibility of estimating the importance of a discovery until the discoverer is dead or too old to enjoy the money. Any jury will inevitably tend to rate the discovery of a fact above the invention of a method of research. Let us take as an example the recent and thoroughly justifiable award of a Nobel prize to Banting and Macleod for the preparation of insulin, the substance by whose injection a victim of diabetes can be restored to health. The numerous extracts prepared before success was reached had all to be tested by injection into dogs in order to study their effect on the amount of sugar in the blood. Now, accurate blood sugar analysis is extremely difficult, especially when one has only a few drops of blood to work on. It has been brought to its present degree of efficiency by some sixty years of very persistent and rather dull work in hundreds of different laboratories. This work had occupied far more time, and probably required far more thought and patience, than the final stage at Toronto. But its value was less obvious at the time and its appeal to the imagination is smaller.

As a matter of fact a great scientist is very lucky if in his own day he receives such recognition as a Nobel prize. Willard Gibbs, the father of modern physical chemistry, was probably the greatest American scientist of the nineteenth century. He was so far recognized by his contemporaries as to be made a professor, though I cannot believe that his lectures were very intelligible. His sister habitually compelled him to drive her round in her buggy, on the ground that her husband was a business man and could not spare the time.

But the fact that under a system of rewards much merit would be unrecognized is the least argument against such a system. No conceivable system can forestall the judgment of posterity It would tend to

divert scientific effort towards the obtaining of sensational rather than solid results. If a prize of a million
dollars had been offered fifty years ago for a substance
whose injection would relieve diabetes (and insulin
would have been very cheap at the price), many of the
men who devised the methods of blood sugar analysis
would have been drawn into fruitless attempts to
isolate insulin

Again, no chemical discovery is more obviously
worthy of recompense than that of a new element.
Three new elements—hafnium, masurium, and rhenium
—have been discovered in the last three years. But
in each case the discovery was made through the
application of Moseley's law connecting the X-ray
spectrum of an element with its atomic number. This
law was arrived at as the result of a series of very careful
and laborious measurements, which furnished the
chemist with a weapon of enormous power. But I do
not suppose that it was considered worthy of a para-
graph in the popular press at the time of its discovery,
and I doubt if the public would stand the allocation of
large sums from its exchequer as a reward for work
which is unintelligible to it.

Scientific discovery should be paid for on a system of
credit rather than of cash down. At present research
in pure science is mainly performed by professors and
lecturers at universities in the intervals of the teaching
and administration for which they are paid. A little
research of this kind is paid for as such by the Medical
Research Council and other public bodies, but the vast
majority of the public money spent on research goes
out for work on various branches of applied science,

such as aeroplane design and practical medicine. The Royal Society has established a few professorships wholly devoted to research work, and one cannot but hope that many more such will be founded. A great discoverer can generally expound his own work, but he may be a thoroughly bad elementary lecturer and an extremely incompetent administrator of a laboratory. It would be an advantage to education as well as to research if these duties could more often be separated.

And everywhere the salaries are extremely low. It is too late to reward Faraday, Hertz, or Pasteur. We can at least see that their successors possess an income large enough to allow them to bring up a family of five children and give them a first-class education, while allowing themselves such luxuries as a small motor car. I can think of no professorship in Britain or France which satisfies this criterion. Our research workers are faced with the choice of deserting their calling on marriage or drastically limiting their families. I can think of some who have become expert witnesses, journalists, civil servants, and even tobacco salesmen, to their own economic advantage, but hardly to that of posterity. Others content themselves with one or two children, a questionable advantage to the public since scientific ability is strongly inherited.

There is, perhaps, a final argument, from the point of view of the scientist, against a system of rewards for research, namely, that it should logically imply a system of punishments for discoveries adjudged to be of disadvantage to the public. The discoverer of a new explosive, a new poison gas, or a new principle in aeronautics, might find himself or herself condemned to the most appalling penalties.   Such discoveries are often made by persons of the mildest character, who have no idea that their work will serve to kill a fly, much less a human being.   And as it is equally true that discoveries of immense general utility are often made by misanthropes only interested in pure theory, and caring not a rap for their own nor any one else's welfare, there is perhaps no injustice in refusing them pecuniary rewards which few of them have ever demanded.

But it is not only unjust but contrary to the public interest that scientific research should be, as it is, the worst paid of all the intellectual professions.

SCIENCE AND POLITICS

FROM time to time I am asked whether I have ever thought of taking to politics. I suppose that question is asked of every man who can speak consecutively for twenty minutes. Sometimes I fear I have answered that politics are no occupation for an honest man. If I made that answer I was wrong, for it is my duty, and every one's duty, to try to alter that state of affairs if it exists. But the true answer was that I thought I could be of more use where I was. ' But why ? ' my questioner might have asked, ' if you can find a method of reducing the amount of potassium in your own blood or altering the distribution of sugar between the different tissues of your body, should you not apply your mind to reducing the amount of unemployment in the country or helping to bring about a juster distribution of its wealth ? ' I could not answer that these questions do not interest me. I have not to take many paces outside my laboratory to see the need for political and social reform. As a skilled manual worker and a trade unionist, I have a strong idea where I should find my political affinities.

I might claim that my work had done something to save life and health in the fight against disease. But if it resulted in halving the death-rate from heart disease (which is highly unlikely) it would not save half as many lives as if I could be instrumental in bringing the sanitary conditions of the unskilled urban labourer up to those of the skilled artisan. And these conditions depend mainly on housing and wages.

My only valid excuse seems to be along quite different lines. I believe that social problems can only be solved in the long run by the application of scientific method, such as has made possible modern industry and modern medicine. I am at once answered by two sets of people. The first tell me that if I think on scientific lines about politics I shall inevitably be led to their own favourite scheme, a scientific tariff perhaps, or a scientific organization of the means of production by the State. The others say that my scientific method may be adequate for dealing with machines or animals ; but that as man is a great deal more than a machine or an animal, it cannot be applied to politics. With these last I have considerable sympathy

If I thought that science in its present embryonic state could be applied to politics I should become a politician. But it certainly cannot. Man is no more a mere animal than he is an economic unit. It is quite true that biological laws apply to him as mechanical laws do. Good intentions alone are as useless against smallpox as against an earthquake, though they are needed for dealing with both these calamities. But to predict the behaviour of men in the mass we require knowledge of a special kind of psychology. And at the present moment the expert politician knows ten times as much of it as the best psychologist. But there is this big difference between the two. What little knowledge the psychologist possesses, though it is so abstract and meagre as to be of very little practical value, can be put in a form accessible to other psychologists. The same cannot be said of the politicians.

Mr. Ramsay MacDonald and Lord Younger disagree on most political topics; but they would probably agree to a large extent in estimating the ability and integrity of a given statesman, or the probability of gaining votes by a given speech or measure. Yet they could not put into words the processes by which they arrive at these estimates, although their judgment is clearly worth more when they agree than when they differ. The psychologists are just beginning to give an account of these processes. In another two or three centuries they will be beating the politicians at their own game and usurping their power, provided that the politicians have left a civilization in which psychology can exist.

I say two or three centuries for the following reason. Two hundred years ago the physicists and chemists were beginning to study the properties of metals by exact methods involving measurement, and the biologists were looking down the first microscopes. But the real knowledge of metals lay in the hands of skilled workmen, who handed down their rule of thumb methods and manual dexterity to their children. To-day metallurgy is a branch of applied science, while biologists are just beginning to be of some use to the practical animal breeder, though they cannot yet beat him at his own game. Psychology is about as much more complex than biology as biology than physics. Hence my estimate of the time it will take to develop. Let us hope that it is too large.

Why then am I not a psychologist? Because, with

all respect to psychologists, I do not think psychology is yet a science.

Mechanics became a science when physicists had decided what they meant by such words as weight, velocity, and force, but not till then. The psychologists are still trying to arrive at a satisfactory terminology for the simplest phenomena that they have to deal with. Until they are clearer as to the exact meaning of the words they use, they can hardly begin to record events on scientific lines. Moreover, I do not believe that psychology will go very far without a satisfactory physiology of the nervous system, any more than physiology could advance until physics and chemistry had developed to a certain point. This is not to say that physiology is a mere branch of physics or chemistry, or the mind a mere by-product of the brain. But it is a fact that we can only know about life by observing the movements of matter.

You may be the most spiritually minded man on earth, but I can only learn that fact by seeing, hearing, or feeling your bodily movements. As the latter depend on events in your brain, I may as well get some information about those events. Moreover, the close dependence of your thought on processes in your brain may readily be shown by comparing the effects on it of a blow on your head and a blow elsewhere. To study psychology before we understand the physiology of the brain is like trying to study physics without a knowledge of mathematics. Physics is more than mathematics, as matter is more than space, but you cannot have the one without the other. Now at the moment the physiology of the nervous system is being worked out with great speed, and by contributing to its progress I suspect that I am doing more for psychology than if I became a psychologist.

It is worth while taking an example of what I mean by the application of psychology to politics. The success or failure of Socialism will depend on whether it can furnish as good an incentive to individual effort as our present economic system, more than on any other single fact. This is a very favourite topic with debating societies. Similarly the question whether the first bird came out of the first egg or the other way round was debated; until a study of fossils showed that birds were descended from reptiles which laid eggs, and therefore the egg had the priority. But a better analogy is the case of human diet. During the war the basis of the people's food had to be changed, and an adequate but not excessive ration assigned. If this had been left to politicians it is fairly certain that while there would have been waste in some directions, yet essential ingredients in the diet would have been left out, and we should have had outbreaks of scurvy, rickets, and dropsy such as occurred on the Continent, even where there was not actual starvation. But the politicians took the advice of some very competent biochemists, and rationing was a success.

A hundred years ago one of Napoleon's armies was put on a cheap and portable diet drawn up by the best physiologists of that day. Some essential ingredients were left out with disastrous results. I fear that the results would be no better if we asked the psychologists of to-day to draw up a scheme of non-economic incentives to effort for a socialized industry. We should do much better to rely on practical experience of the fighting services, municipal enterprises, labour battalions in the war, and so on. But even a hundred years hence I think the psychologists might be quite as valuable as the ' practical men.'

For the moment, then, I believe that the man with a gift for thought on scientific lines is of more use to his fellows in the laboratory than out of it. He can work on clearly defined problems to which his really accurate if rather narrow type of thought can be applied. I know that this is little comfort to the unemployed workman or the war widow who watches the approach of the conflict that will claim her only son. But we have not yet got the general principles to apply to their problems, though we may applaud and support efforts to remedy individual evils. To take a problem which is much nearer solution, that of cancer, we can tell the sufferer to go to the surgeon, who will give him a sporting chance of recovery if the disease is not too far advanced. (The surgeons can cure about fifty per cent. of cases of cancer of the breast or womb if they come in time.)

But the surgeon's methods are rough and often ineffective, though we have every reason to think that in fifty years we shall be able to deal with this disease as we have with typhoid and smallpox. However, few men who are dying of inoperable cancer will be cheered by the news that cancerous tissues have been shown to be capable of breaking up sugars far more rapidly than ordinary tissues can, though this may be the clue to the ultimate prevention or cure of cancer.

To say that the scientific mind is still best employed outside politics, is not, of course, an assertion that

politics are only suited for fools. If I may conclude
with some remarks on a subject of which I know no
more than it is my duty as a voter to know, I would
suggest that with the extension of the State's activities
the organizing type of mind is becoming more necessary
in politics than the emotional and sympathetic type
which has led great causes in the past. When this
country makes its first experiments in Socialism we shall
need in control of the nationalized industries the type of
mind that has from time to time successfully reorgan-
ized the fighting services in the past. It will be the
duty of the Government to entrust them to such a person
though he or she were as tongue-tied as Sir Eric Geddes,
and as adulterous as Samuel Pepys. Fortunately such
a man or woman is quite as likely to rise to power by
organizing a great trade union, as by the paths which
lead to success in the older parties. But as long as the
principles of politics are unsystematized and incom-
municable, I, for one, shall continue to regard any
political projects as interesting experiments which may
or may not promote human happiness but will certainly
furnish important data for future use.

Of the biochemical experiments which I have done
on myself, perhaps two-thirds have had the results for
which I was looking. But every one has also had
unexpected, and many of them unpleasant, effects.
Nothing can be more instructive than to read the fore-
casts which the more intelligent economists of the
eighteenth century, Adam Smith in particular, made as
to the effects of the economic system which was substantially adopted in the nineteenth. Smith foresaw
the great increase of wealth and education. He did

not predict the amazingly unequal distribution of incomes, the periodic waves of unemployment, the gigantic industrial conflicts.

And until politics are a branch of science we shall do well to regard political and social reforms as experiments rather than short cuts to the millennium. The time-scale of evolution so far has been the geological time-scale, in which we expect no substantial change in less than a hundred thousand years. We may be thankful if we have speeded up this progress a thousandfold and at the end of a long life can leave the world noticeably better than we found it.

EUGENICS AND SOCIAL REFORM

PERHAPS the greatest tragedy of our age is the misapplication of science. It is notorious that the principal result of many increases in human power and knowledge has been either an improvement in methods of destroying human life and property, or an accentuation of economic inequality. This is largely the fault of the confused thinking of 'advanced' politicians. I refer to mental processes such as that which led to our forgoing the use of 'mustard gas,' the most humane weapon ever invented, since of the casualties it caused, 2·6 per cent. died, and $\frac{1}{4}$ per cent. were permanently incapacitated. No one at Washington even suggested abandoning H.E. and shrapnel, which kill or maim about half their casualties. Save for the fact that it was only preventive medicine which rendered practicable the large concentrations of troops in the Great War, biological progress has been little exploited for human hurt. But the growing science of heredity is being used in this country to support the political opinions of the extreme right, and in America by some of the most ferocious enemies of human liberty. And yet it seems likely that the facts, in so far as they are applicable to politics at all, would warrant conclusions of an entirely different nature from those which have so far been drawn, and which have made eugenics abhorrent to many democrats.

The relevant facts fall into two classes, first those

190

which relate to hereditary abnormalities or tendencies to disease, and secondly those bearing on the inheritance of intelligence and the different birth-rate in different social classes. The former are comparatively unimportant. Some, like certain types of eye malady, are transmitted only by affected people, and to about half their children. It is on the whole undesirable that they should beget their like ; but before we begin curtailing the liberties of people already sufficiently unfortunate, we should first try to impress on them their duty to restrict their families, and to see that they have the means to do so.

Other hereditary troubles, apparently including much feeble-mindedness, are mainly transmitted by unaffected people, and not necessarily by affected. If every feeble-minded person were sterilized, hundreds of generations would be needed to eliminate hereditary feeble-mindedness. The question of whether a feeble-minded girl should be allowed to produce an indefinite number of bastards ought to be settled on the same grounds as any other social problem. Such a woman is quite as likely to harm her contemporaries by transmitting venereal disease, and her children by negligence, as to be responsible for the idiocy of future generations. And in particular any legislation which does not purport to apply, and is not actually applied (a very different thing), to all social classes alike, will probably be applied unjustly to the poor

But hereditary deaf-mutism and feeble-mindedness, though serious evils, are not menaces to national existence, and it is claimed that this is the case with the differential birth-rate in different classes. We must first examine the question how far heredity rather than environment is responsible for the mental differences between the children of different social classes. The question cannot be answered on *a priori* grounds. To take a simple case, illiteracy in England is mainly determined by congenital weak-mindedness, in India by parental poverty. The problem has been attacked along many lines. For example, tests such as the memorizing of arbitrary forms, or the time taken to sort out cards of five different colours from a mixture, showed as great differences between the children of members of the professional classes and the equally well-fed and healthy children of middle-class parents of similar incomes, as between the latter and the children of the poor. Again, within schools drawn from a fairly homogeneous population the correlation between brothers or sisters with regard to intellectual performance was just the same as that with regard to eye-colour, which certainly does not depend on environment. Finally, we have the cases of ' identical ' twins, who have the same hereditary make-up, and are almost or quite indistinguishable physically. Perhaps the best example of the extraordinary similarity of their mental processes is the well-authenticated story of the pair who were brought before their headmaster for making the same mistake in a mathematical examination, and were able to prove that they had been in different rooms at the time !

All investigators are agreed that mental capacity is strongly hereditary, though, as with stature, environment plays a part in its determination. Of course, two fools may produce a genius, or two dwarfs a giant, but such cases are the exception. It is also agreed that among the poorer nine-tenths of the population the abler members on the whole tend to rise into a richer class than their parents. This is not so among the rich, where the more intelligent are commonly content with incomes of £1000-£2000; though even here the converse holds, and fools and their money are soon parted. Finally, there is no doubt that the richer classes breed much more slowly than the poorer, and that this is not compensated for by lower infantile mortality. A thousand married teachers under fifty-five annually beget [1] 95 children, a thousand doctors 103, carpenters 150, general labourers 267. Thus the unskilled workers are breeding much faster than the skilled classes, and, in view of the demands for intellectual and manual skill in modern civilization, this is an evil. The Eugenics Education Society have doubtless done good work in persuading a certain number of intelligent people that it is their duty to have more children. They have also rightly urged lessened taxation of parents of children. But many of their members have coupled this with a clamour against measures designed to ameliorate the lot of the children of the poor at the expense of the rich. It is a curious policy to combat evils due to economic inequality by perpetuating that inequality.

The main reasons for the differential birth-rate seem to be three. First, persons who have money to spend on the education of their children and a prospect of providing them with substantial sums as legacies, marriage settlements, and the like, restrict their families in the interests of individual children. It

[1] These are pre-war figures, now greatly reduced.

therefore follows that any measures which tend to disseminate heritable property among the poor, such as the breaking up of large estates, are eugenically desirable. So are drastic improvements in our elementary education and in our scholarship system. The average doctor would probably beget at least one more child if he could be sure that his children would be satisfactorily educated at State or State-aided schools, and that there were sufficient scholarships available to enable any child of intelligence appreciably above the average to enjoy a cheap university education. If I attached the importance to eugenics which certain people claim to do, I should, I think, be bound to advocate the complete abolition of hereditary property, and the free and compulsory attendance of all children at State schools. At any rate, all legislation tending in these directions must be regarded as eugenic.

Secondly, the poor do not know how to restrict their families by artificial means, and their overcrowding renders marital continence and even preventive measures very difficult. The eugenist must therefore approve of better housing schemes, and of all movements designed to spread a knowledge of birth-control among the poor. Thirdly, a certain section of the poor are extremely improvident, and do not consider the consequences of their actions. This leads to poverty on the one hand and large families on the other. But this section probably do not possess any great desire for children, which is, after all, one of the most respectable and unselfish of the elementary human desires. If a knowledge of birth-control were universally diffused I am inclined to believe that they would produce considerably less children than their more provident neighbours, who mostly possess much stronger parental instincts.

Finally, there is a group of causes of a more subtle nature. For example, the women of the richer classes probably suffer unduly in childbirth from lack of previous exercise of a suitable character, and also tend to restrict their families on account of other competing interests. In other words, rich women need more exercise, and poor women more education. Again, those who have risen in the economic scale by their intelligence are often cut off by differences of tradition from partners of the class into which they have risen, by differences of intelligence and education from that of their parents. Class-consciousness is dysgenic in a society possessing a social ladder.

It will be seen, then, that the differential birth-rate is very largely determined by social inequalities of a type already recognized and deplored. If these were remedied the main characters favoured by selection would be health and strong parental instincts instead of—as now—the type of mental equipment which prevents a man from becoming well-to-do.

Just the same argument applies to racial problems. It was only the emancipation of the negroes which saved the United States from twice its present black population. This event gave them access to alcohol, venereal diseases, and consumption. Their rate of increase slowed down at once, and it is only between the last two censuses that the absolute excess of negro births over deaths has once more equalled that in the decade before the Civil War. In the Northern States and in all towns the negro death-rate exceeds the birth-rate. If in the interests of racial purity all negroes were expelled from north of the Mason-Dixon line the proportion of blacks to whites in the whole Union would be markedly increased within a generation. And prohibition has probably been of far greater benefit to blacks than whites. Those who are convinced of the superiority of Europeans to Indians may console themselves with the thought that a British withdrawal from India would cause a very rapid decrease by war and famine in the number of Indians, and remove any danger of the Indianization of other parts of the Empire.

To sum up, the rational programme for a eugenist is as follows : Teach voluntary eugenics by all means ; but if you desire to check the increase of any population or section of a population, either massacre it or force upon it the greatest practicable amount of liberty, education, and wealth. Civilization stands in real danger from over-production of 'undermen.' But if it perishes from this cause it will be because its governing class cared more for wealth than for justice.

OCCUPATIONAL MORTALITY

FEW Government documents are more interesting than the Registrar-General's decennial report on occupational mortality in England and Wales, and it is safe to say that none of equal importance is more neglected. This is partly due to the fact that it refers to the years 1910-1912, and is therefore already out of date, partly perhaps because most of it is unsuited for party propaganda.

The figures given contain a few pitfalls, most of which are pointed out in the introduction. New trades, for example, such as electricity supply and motor-driving, attract young men, and therefore have spuriously low death-rates. Sailors are absent when the census is taken, but come home to die, and therefore seem to be very unhealthy. Navvies apparently describe themselves as such at the census, but their relatives tell the Registrar of Deaths that they were general labourers. They thus appear to be healthier than is the case, and general labourers less so. And hawkers, whose death-rate is nearly double the average, are unhealthy, not so much on account of the conditions of their work, as because the failures in other occupations drift into this calling in the last years of their life.

When allowance is made for these facts, the high mortality in those callings where it exceeds the average by 50 per cent. or more is due to two, and only two, causes—alcohol and dust. The most dangerous of all occupations, with a death-rate almost two and a half times the average, is that of barman. The guilt of this death-rate is, however, about equally divided between the manufacturers of strong drink and the advocates of temperance. As compared with their landlords, barmen are less than half as likely to die of alcoholism and its sequel, cirrhosis of the liver; but they are more than twice as likely to perish of consumption and other lung diseases, which, indeed, account for nearly half their mortality. These diseases are mainly due to the overcrowding and under-ventilation of their places of work. Though the health of barmen is doubtless better since their hours of work have been restricted, it will continue to be bad as long as bars are so few as to be overcrowded, and so dark and close as to form ideal breeding-grounds for the tubercle bacillus. But any serious attempts to bring the sanitary conditions of the best bar up to those of the worst factory are resisted in the interests of temperance as likely to render public-houses more attractive. Any Government which took health seriously would either abolish our bars or ventilate them.

The next most unhealthy trades, with a death-rate just double the average, are tin-miners and file-makers. In each case the main cause of death is consumption, due to the clogging of the lungs with dust. Fortunately, only a few kinds of dust are harmful, and some seem to be beneficial. Coal-miners' lungs are perfectly black, but their death-rate from phthisis is half that of the general population, and the same as that of agricultural labourers. A great deal has been done of late years to reduce the danger to tin-miners by laying the rock-dust with spray, and it may be hoped that the results of this will be apparent in the statistics of 1920-1922. The death-rate of innkeepers is 60 per cent. above the average.

Urban, though not rural, poverty comes third on the killing list—first, if we consider the numbers involved. Unskilled urban labourers have a death-rate 43 per cent. above the average ; and their death-rate is higher than the mean from every single cause except lead poisoning, a disease of skilled labour which kills less than a hundred men annually, and diabetes, which is largely due to over-eating. If their conditions could be brought up to the average (not a very ambitious ideal) we could save thirty thousand lives of adult males a year, not to mention women and children, as good a result as if we had abolished heart disease or pneumonia.

At the other end of the scale, the healthiest occupation appears to be the manufacture of glue and manure, an interesting commentary on the widely held theory that bad smells cause bad health. These men, with electricity supply workers and machine compositors (though not ordinary printers), alone have less than half the average death-rate ; and even if the figures can to some extent be explained away, these trades must be conspicuously healthy. They are followed in the scale of health by clergymen, gardeners, farm-labourers, and gamekeepers in that order. The other intellectual callings, such as teachers, doctors, and lawyers, are all healthier than the average. But it is an interesting and rather beautiful fact that, while doctors have the highest mortality of this group, their children have just half the death-rate of those of any other large profession. The next healthiest indoor trades are lithography and soap-making, though the latter may owe its position rather to the personality of Lord Leverhulme than to its intrinsic nature.

Apart from murder, the most interesting though not the most important causes of death are suicide and accident. Barmen head the suicide list, being closely followed by chemists, hairdressers, and innkeepers, while doctors and cutlers are among the more suicidal callings. Alcohol and opportunity thus seem to determine suicide, but why should tin and lead miners utterly refuse to commit this rash act ? *Spes phthisica*, perhaps ? In spite of the fact that most of their violent deaths take place on the high seas and are therefore not registered, seamen head the accidental death list with four times the average rate, though bargees and lightermen almost equal them. If these latter were all taught to swim, their mortality from accident could probably be reduced to that of dockers at least. A philanthropist who wished to save life could hardly do it more cheaply than by offering £5 to every bargee who could pass a swimming test. Coal-miners have a death-rate from accident of only a little over double the average ; and this has probably been considerably reduced since the use of stone dust has abolished large colliery explosions. Even when their deaths from accident are included, coal-miners live longer than most people ; and when deaths from disease only are considered, they are as healthy as lawyers.

In all but a few occupations the greatest killer, and one which is the more serious because it kills in what should be the prime of life, is phthisis. Apart from barmen and hawkers, all the phthisical trades are dusty.

Practically every man who works a machine-drill in a tin-mine or in the Transvaal dies of consumption. Out of 142 men in the Redruth registration district of Cornwall who had worked machine-drills, all but nine died of lung disease. Tin-miners as a whole have about five times the average mortality from phthisis; cutlers, barmen, file-makers, and stonemasons working in sandstone three times; lead-miners and potters twice; while clergymen, gamekeepers, and locomotive drivers have about a third.

In view of the large amount of guessing which goes on as to the cause of cancer, its occupational incidence is interesting. It varies far less from one trade to another than that of most diseases; so, as it is far commoner in civilized than uncivilized countries, its main cause must be sought in some condition common to all walks of civilized life. It is not perhaps without significance that barmen and brewers have the highest cancer rates, though it must be admitted that innkeepers are only slightly above the average. But it is consoling to find that tobacco manufacturers have the lowest cancer mortality of any trade, while tobacconists are also unusually immune. Chimney-sweeps used to die in large numbers from cancer due to irritation of the skin by soot; but while the cancer rate for most occupations has risen, theirs has fallen considerably, apparently through increased cleanliness.

An interesting light is thrown on chastity by the death-rates from locomotor ataxy and general paralysis, which are the results of syphilis. Seamen, bargees, and waiters have more than twice the average mortality from these complaints, while fishermen and commercial

travellers run them close. Farmers, clergymen, and
gamekeepers are the chastest of men. The richer
classes and unskilled labourers are less chaste than the
mass of the people.

Some of the conclusions to be drawn are to my mind
sufficiently clear. It is our plain duty to deal drastically with those occupations in which harmful dust is
inhaled. We tax alcoholic drinks on the ground that
they harm certain people, although Pearl has shown
that moderate but habitual drinkers are distinctly
healthier than total abstainers. But assistance from
the State in the form of a tariff is actually obtained by
the cutlery trade, which has three times the average
death-rate from consumption. Until such trades mend
their ways, it is the clear duty of the Government to
discourage them, and private individuals can play their
part. It would appear, for example, to be a patriotic
duty to buy British food, leather, and coal in preference
to foreign, as agriculture, tanning, and coal-mining
are healthy occupations ; but if so, it is an equally clear
duty to buy foreign scissors, files, and pottery. A great
deal has been done to reduce the dust and consumption
in the cutlery trades by wet grinding and ventilation.
This is relatively easy in factories, but unfortunately
a large proportion of files, for example, are made by
workmen in their own houses and backyards. This
type of industry is dear to the hearts of our mediaevalists,
not to mention the late Prince Kropotkin, but it does
not take kindly to sanitary regulations. An enlightened
public opinion could wipe out industrial phthisis as it
has wiped out lead and phosphorus poisoning. So long,
however, as one death from avoidable accident occupies

as much space in the Press, and therefore in the public mind, as ten thousand from avoidable consumption, such a result is hardly probable. But the Government might at least contribute to such an end by issuing the information on this topic at a price of less than 13s. 6d., and after a delay of less than ten years.

WHEN I AM DEAD [1]

WHEN I am dead I propose to be dissected; in fact, a distinguished anatomist has already been promised my head should he survive me. I hope that I have been of some use to my fellow creatures while alive, and see no reason why I should not continue to be so when dead. I admit, however, that if funerals gave as much pleasure to the living in England as they do in Scotland I might change my mind on the subject.

But shall I be there to attend my dissection or to haunt my next-of-kin if he or she forbids it ? Indeed, will anything of me but my body, other men's memory of me, and the results of my life, survive my death ? Certainly I cannot deny the possibility, but at no period of my life—least of all during the war, when I was nearest to death—has my personal survival seemed to me at all a probable contingency.

If I die as most people die, I shall gradually lose my intellectual faculties, my senses will fail, and I shall become unconscious. And then I am asked to believe that I shall suddenly wake up to a vivid consciousness in hell, heaven, purgatory, or some other state of existence.

Now, I have lost consciousness from blows on the head, from fever, anaesthetics, want of oxygen, and other causes ; and I therefore know that my consciousness

---

[1] This paper was one of a series on the above topic, most of which supported a different thesis.

204

depends on the physical and chemical condition of my
brain, and that very small changes in that organ will
modify or destroy it.

But I am asked to believe that my mind will continue
without a brain, or will be miraculously provided with
a new one. I am asked to believe such an improbable
theory on three grounds.

The first set of arguments are religious. My
Catholic friends hope to survive death on the authority
of the Church ; some of my Protestant acquaintances
rely on the testimony of the Bible. But they do not
convince me, for the Church has taught doctrines which
I know to be false, and the Bible contains statements—
for example, concerning the earth's past—which I also
know to be false.

Other Christians ask me to believe in immortality on
the authority of Jesus. That is a much more cogent
argument, because Jesus was not only a great man and
a great ethical teacher, but a great psychologist. But
the characteristic part of any man's teaching is what is
novel and heretical in it, and not what he and his
audience take for granted.

Now Jesus, and the Pharisees, at any rate, among his
hearers, took the resurrection of the dead for granted.
They also took for granted that madness was due to
possession by devils.

When Jesus tells me to love my enemies he is speaking
his own mind, and I am prepared to make the attempt ;
when he tells me that I shall rise from the dead he is
only speaking for his age, and his words no more convince me of immortality than of demoniacal possession.

Again, I am told that men have always believed in

immortality, and that religion and morality are impossible without it.   The truth of this ludicrous statement may be tested by referring to the first seven books of the Bible.   They are full of religion and ethics, but contain no reference to human survival of death.   Nor did the Psalmist believe in it.   ' The dead praise not Thee, O Lord,' he said, ' neither all they that go down into silence.'

As a matter of fact, the belief in a life beyond the grave reached its culminating point in Egypt four or five thousand years ago, when the rich, at any rate, seem to have spent more money in provision for their future life than for their present.   To judge from what has come down to us of his writings, Moses, the Man of God, who was well versed in Egyptian religion, had no more use for a future life than for the worship of crocodiles.

The belief in personal immortality seems to have spread gradually out from Egypt, along with the use of copper, bronze, and gold, and often, especially in Polynesia, accompanied by the practice of mummification. It is an attractive doctrine, and is only now beginning to lose its grip on the human mind.   But it is losing it.

Most of the men under my command whom I got to know during the war believed in God.   But I think the majority thought that death would probably be the end of them, and I am absolutely certain that that is the view of most highly educated people.

I am also asked to believe in a future life on philosophical grounds.   There are a number of arguments which seem to prove that my soul is eternal and indestructible.   Unfortunately, they also prove that it has existed from all eternity. And, though I am quite willing to believe that ten years after my death I shall be as I was ten years before my birth, the prospect cannot be said to thrill me.

Again, it is argued that justice demands a future life in which sin will be punished, virtue rewarded, and undeserved suffering compensated. Such a view would, of course, involve a future life for animals, in which the hunted hare would demand of its pursuers for what crime it was torn to pieces.

But it assumes that the universe is governed according to human ideas of justice. The sample of it which we know is certainly not so governed ; and I see no reason to suppose that its inequalities are redressed elsewhere. No doubt I should like to see them redressed, but then I should like to see England a land fit for heroes to live in, though I do not suppose that I will.

As a matter of fact, conditions in the present world have been improved largely by recognizing that the laws governing it are not the laws of justice but the laws of physics. As long as people thought that cholera epidemics were a punishment for the people's sins they continued. When it was found that they were due to a microbe they were stopped.

We do ourselves no good in the long run by telling ourselves pleasant fairy tales about this world or the next. If we devoted the energy that we waste in preparing for a future life to preventing war, poverty, and disease, we could at least make our present lives very satisfactory for most people, and if we were happy in this world we should not feel the need of happiness hereafter.

It is worth remembering how very few people in the past have believed in the justice of the universe. Most Christians have believed that unbelievers and unbaptized children were doomed to spend eternity in hell, while Buddhists believe that all existence is an evil. It is only our own perhaps unduly optimistic age that has assumed that a future life is to be desired.

The arguments of the spiritualists, theosophists, and their like claim to be based on evidence of a kind which appeals to the scientific mind, and a dozen or so distinguished men of science have been spiritualists. Some of this evidence is based on fraud. The bulk of it proves nothing. I have often taken part in the receipt of messages alleged to come from spirits, but they have never given any verifiable information unknown to any members of the circle.

Even, however, if we accepted the view of the spiritualists that a medium can somehow get into communication with the mind of a dead man, what would this prove ? If we accept spiritualism we must certainly accept telepathy. Now, I can see little more difficulty in two minds communicating across time than across space.

If I can transmit thoughts to a friend in Australia to-day, that does not prove that my mind is in Australia. If I give information to a medium in the year 1990, ten years after my death, that will not prove that my mind will still be in existence in 1990.

To prove the survival of the mind or soul as something living and active we should need evidence that it is still developing, thinking, and willing ; spiritualism does not give us this evidence. Shelley is said to have dictated a poem to a medium. It was a very bad poem.

Nor do the post-mortem productions of Oscar Wilde reach the standard which he attained when alive.

The accounts given by spirits of a future life vary from land to land and from age to age. Mediaeval ghosts generally come from purgatory, like Hamlet's father ; more rarely from heaven or hell. Hindu and Buddhist ghosts are awaiting their next incarnation. Modern European spirits usually profess a rather diluted Christianity. In fact, the evidence as to the nature of a future life is so contradictory that we must in any case reject most of it.

Personally, I think that all accounts of a future life are mere reflections of the medium's own opinions on the subject, which are of no more value than any one else's. (The possible interpretation of spiritualistic phenomena which I have given is one which has commended itself to some of their most careful investigators, but has obtained little publicity as it has no emotional appeal.)

But if death will probably be the end of me as a finite individual mind, that does not mean that it will be the end of me altogether. It seems to me immensely unlikely that mind is a mere by-product of matter.

For if my mental processes are determined wholly by the motions of atoms in my brain I have no reason to suppose that my beliefs are true. They may be sound chemically, but that does not make them sound logically. And hence I have no reason for supposing my brain to be composed of atoms.

In order to escape from this necessity of sawing away the branch on which I am sitting, so to speak, I am compelled to believe that mind is not wholly conditioned by matter. But as regards my own very finite and imperfect mind, I can see, by studying the effects on it of drugs, alcohol, disease, and so on, that its limitations are largely at least due to my body.

Without that body it may perish altogether, but it seems to me quite as probable that it will lose its limitations and be merged into an infinite mind or something analogous to a mind which I have reason to suspect probably exists behind nature. How this might be accomplished I have no idea.

But I notice that when I think logically and scientifically or act morally my thoughts and actions cease to be characteristic of myself, and are those of any intelligent or moral being in the same position ; in fact, I am already identifying my mind with an absolute or unconditioned mind.

Only in so far as I do this can I see any probability of my survival, and the more I do so the less I am interested in my private affairs and the less desire do I feel for personal immortality. The belief in my own eternity seems to me indeed to be a piece of unwarranted self-glorification, and the desire for it a gross concession to selfishness.

In so far as I set my heart on things that will not perish with me, I automatically remove the sting from my death. I am far more interested in the problems of biochemistry than in the question of what, if anything, will happen to me when I am dead.

Until this attitude is more general the latter question will remain too charged with emotion to make a scientific investigation of it possible. And until such an investigation is possible a man who is honest with himself can only answer, ' I do not know.'

THE DUTY OF DOUBT

W E are taught that faith is a virtue. This is obviously true in some cases, and to my mind equally false in others. There are occasions when the need for it must be emphasized. Nevertheless, at the present time I believe that mankind is suffering from too much, rather than too little faith, and it is doubt rather than faith that must be preached. I am not thinking wholly or even mainly of faith in the Christian or any other religion, but simply of the habit of taking things for granted. Nor am I praising a blind and haphazard doubt, which is as unintelligent as blind faith, and far less fruitful. Greece and Rome produced a sect of sceptic philosophers who gave valid reasons for doubting anything whatever, and finally left themselves with no motives except the gratification of their instincts. Christianity swept away scepticism along with many nobler philosophies. And any system in which the suspense of judgment leads to the suspense of action will inevitably perish at the hands of men who are prepared to act, however utterly nonsensical be the motives that lead them.

Modern science began with great acts of doubt. Copernicus doubted that the sun went round the earth, Galileo that heavy bodies fall faster than light ones, Harvey that the blood flowed into the tissues through the veins. They had each a theory to replace the old one, and their observations and experiments were largely designed to support that theory. But as time went on these theories, too, were found wanting. The planets do not go round the sun in circles as Copernicus thought ; gravitation is a more complex affair than Galileo or even Newton believed. And nowadays, though many experiments are made to support old or new theories, large numbers merely go to prove them false without putting anything in their place. One can hardly open a scientific journal without finding a paper with some such title as ' On an Anomalous Type of Inheritance in Potatoes,' or, ' Deviations from the Law of Mass Action in Concentrated Sugar Solutions.' The statement of any general principle is enough to raise active doubt in many minds. Moreover, the authors very often make no attempt to put forward an improved theory ; and if they do so it is generally in a very tentative form. ' The results so far obtained are consistent with the view that . . . ' has taken the place of ' Thus saith the Lord . . . ' as an introduction to a new theory. Moses apparently regarded 'An eye for an eye and a tooth for a tooth ' as an absolute principle of right conduct ; Einstein would certainly not regard any of his laws as final accounts of the behaviour of matter.

Now, the method of science, which involves doubt, has been conspicuously successful over a certain field. But there are many who affirm that that field is strictly limited. ' In the realm of religion and ethics,' they assert, ' we have reached finality. You may not be certain about the principles of physics, but I and every right-minded man and woman are certain about the principles of right and wrong ; and those who question them deserve to be treated as criminals.' This attitude is rather commoner in the United States than in most civilized countries, not because Americans are more stupid or less educated than other nations, but because they live amid a more homogeneous moral tradition. The Englishman who thinks it wrong to live with a mistress has only to cross to France to find people doing so without exciting serious disapproval. The Russian who regards making a fortune as a disgusting vice has only to enter Finland (if his Government will let him) to find quite decent and useful individuals practising it. But the American has a long way to travel before he or she will find otherwise respectable women smoking cigars without exciting unfavourable comment, or governing classes who regard the self-made millionaire as inevitably vulgar and unpleasant.

Now, there are conditions under which it is an advantage that moral principles should be unquestioned. It is roughly true that our laws are the laws which would have been suitable for our grandparents, and our moral code that which would have sufficed for our great-great-grandparents. It takes about two generations of effort to effect a great legal change, say Prohibition or Irish Home Rule, and a good deal more to dethrone a generally accepted principle of moral conduct, such as the different moral standards of the sexes or the wickedness of sport on Sundays. In a society which is not altering much in other respects this stability is an excellent thing, though of course the desirable moral code will vary from place to place. Thus the South Pacific islanders almost universally practised infanticide or abortion, and very often cannibalism or head-hunting.

The islands were as thickly populated as was possible with the methods of agriculture and fishing available, and if the population had not been kept down by these methods famines would have occurred. The missionaries have taught them that these practices are wrong; and so they are now, since European diseases and drinks have replaced them as checks on over-population.

Now, the moral code of Europe, North America, Australia, and New Zealand is to a large extent the code which was found to work in mediaeval Europe. Of course, it has altered since the Middle Ages, but it is far more similar to its ancestor of six hundred years back than to the codes, say, of China, Arabia, New Guinea, or Central Africa to-day. The mediaeval code was evolved in a society mainly engaged in small-scale agriculture and small-scale industry, dominated by a small educated class of priests, and a still smaller military nobility. And the oddest traces of this survive even in the United States. University professors are no longer in holy orders, but they are expected to conform to a standard of conduct much stricter than that demanded of business men or soldiers. The head of the state no longer wears a sword and chain-mail on public occasions. (I am not talking about kings, who still occasionally wear swords, and who, when explosives have been superseded by other methods of killing, will probably carry dummy bombs.) But he still behaves to the heads of other states in a manner appropriate to a mediaeval knight. We are delighted (at least if we are shareholders) when company presidents and directors effect a combination with another corporation in the same line of business, but we expect our premiers and presidents to maintain our national independence to the last drop of our blood.

And the same applies to property. It was obviously right that a mediaeval workman should own his own tools and workshop. It is obviously impracticable for a modern factory worker to own half a lathe and twenty square yards of floor space. It is only gradually being realized that the idea of absolute personal ownership so suitable when applied to a spade or a chisel leads to inconveniences when applied to a share certificate. And those who realize it most fully are convinced—why, I am not very clear—that those inconveniences would vanish if only the ownership were transferred to the State. The truth is more probably that the idea of absolute ownership is ceasing to work and will have to be replaced, as the idea of absolute position has been in physics or that of fixed species in biology. The believer in absolute ownership will at once ask me what I have to put in its place, and will raise a triumphant shout when I say that I do not know.

Now, supposing I go to a physiologist and convince him that his otherwise admirable theory of conduction in nerves will not explain, let us say, the action of cocaine in blocking them, he will not immediately ask me to produce a theory better than his own. Nor will he abandon his former view ; he will try out modifications of it and see whether they work. He will quite probably spend a couple of months in experiments suggested by a theory which he regards as likely to be false. And when he arrives at a scheme of ideas which will fit all the facts so far known he will hardly dignify it by the name of theory, but call it a working hypothesis.

'Yes,' my opponent will say; 'and do you expect men to die for a working hypothesis as they will die for a faith ? '

Well, men have died for odder things. On the occasion of Napoleon III.'s *coup d'état* in 1851, Baudin, a deputy of the Second Republic, was trying to rally opposition in the streets of Paris, though with little hope of success. A workman shouted, ' Why should we risk our lives for your twenty-five francs ? ' referring to his daily salary as a deputy. ' Stay here,' said Baudin, ' and you shall see how a man dies for twenty-five francs.' He died.

And every day men do risk their lives for working hypotheses. Half the art of war consists of doing so. The dispositions of the enemy during a modern battle are more or less unknown. On the available evidence the commander-in-chief forms a hypothesis on which he must then act with the utmost vigour. The great general is the man who stakes everything on his hypothesis while realizing that it is only a hypothesis and must be modified from moment to moment.

Just the same is true of scientific work. A good many biologists experiment on themselves. Of course, it is occasionally necessary to make experiments which one knows are dangerous, for example in determining how a disease is transmitted. A number of people have died in this way, and it is to my mind the ideal way of dying. Others make experiments which are apparently risky, but really perfectly safe provided the theory on which they are based is sound. I have occasionally made experiments of this kind, and if I had died in the course of one I should, while dying, have regarded

myself not as a martyr but as a fool. For all that, I have
no doubt that the theories to which I entrusted my life
were more or less incorrect. One at least has already
been proved so, and the history of science makes it
clear enough that many of the others will be. But
though they had their flaws, they were good enough to
enable me to predict the safety of those particular
experiments, and I hope that I never regarded them as
much more than working hypotheses.

My objection to the thought of many people on all
subjects, and of all people (including myself) on some
subjects, is that it is in a pre-scientific stage. They
seem to be incapable of acting on certain momentous
topics unless they are certain of their premises. Now,
all I should be prepared to say in favour of democracy
is that it is, in my opinion, the least objectionable form
of government so far devised for men and women of
certain sections of the human race. But acting on that
opinion I should be willing to risk my life on its behalf
in defending it against government by a military auto-
crat like the Kaiser or a secret society like the Ku Klux
Klan. Yet I hope that I have not closed my mind to
the claims of other forms of government, for example
the rule of such a voluntary aristocracy as the governing
group of Italy or Russia.

Similarly, in the field of religion it seems to me very
probable that in certain respects the structure of the
universe resembles that of my own mind. This
opinion leads, I think, to implications as to moral conduct different from those of materialism. But if we
try to clothe this idea in the terminology of religion we
can do it in many different ways. Some of these may

serve to make man more like God ; they also have the converse effect in bringing God, in our ideas at least, down to the level of man.

It is characteristic of a good scientific theory that it makes no more assumptions than are needed to explain the facts under consideration and predict a few more. For example, it is quite likely that the inverse square law describing the force between two electrically charged bodies ceases to hold when they are very close or very far apart.   In half an hour I could write down a dozen laws of a more complicated kind which would agree equally well with all the observed facts.   But no one nowadays would be interested in such a law.   Scientific men agree to suspend judgment when they do not know. On the whole, however, the opposite has been the case in the history of religion.   Where there was obvious room for different opinions, for example as to the nature of Jesus' relationship with God, a highly complex theory was gradually built up and was accepted by most Christian churches.   The Unitarians regard themselves as more reasonable than the Trinitarians and have adopted a quite different theory.   To my mind a far more rational view than either would be as follows : ' I believe in God and try to obey and imitate Jesus, but I do not know exactly what is their relationship.'   That is certainly the view of millions of Christians, but no important religious body dares to adopt it. They prefer to go on thinking along pre-scientific lines. And it is this pre-scientific outlook of religion, rather than anything specific in its tenets, which brings it into conflict with science.   ' A creed in harmony with the thought of to-day ' is no better than the Athanasian

Creed if it is taken as a creed and not a working hypothesis, for the simple reason that it will not be in harmony with the thought of to-morrow.

As a matter of fact, the Christian attitude to faith probably rests on a misunderstanding. Diseases of the nervous system and chronic diseases of the skin are particularly amenable to cure by suggestion and other psychological methods. Jesus' recorded healing work was mainly confined to these complaints, and required faith in the patients. But this faith was a belief that they would be cured, and not an assent to historical or metaphysical propositions. Christian Science is so often therapeutically successful because it lays stress on the patient's believing in his or her own health rather than in Noah's Ark or the Ascension. But the Christian churches have tended to accumulate more and more dogmas in their schedules as time went on, so that faith has become more and more intellectual and more and more of a strain on the intellect.

It is just the same with politics. Political creeds fall into two classes. There are the conservative beliefs that institutions which have worked fairly well in the past will go on working under new conditions. Opposed to them are the radical beliefs that policies which have not been tried at all, such as universal disarmament, or have been tried far away or long ago, for example Prohibition in Arabia, are the only solutions for our problems. The good party men honestly hold these beliefs ; the politicians say that they hold them. Fortunately, this is rarely the case, though occasionally an honest man like Robespierre or W. J. Bryan rises to power and acts as if he believed in his own speeches. As long as the average voter's thought is pre-scientific, a politican dare not say : ' I am inclined to think the tariff on imported glass should be raised. I am not sure if this is a sound policy ; however, I am going to try it. After two years, if I do not find its results satisfactory, I shall certainly press for its reduction or even removal.'

Nevertheless, the successful politician often acts in very much that way, and quite calmly goes back on his policy of a year ago. His enemies accuse him of broken pledges ; his friends describe him as an inspired opportunist. In England and the United States the two-party system permits a government to remedy the grosser mistakes of its predecessors, while continuing their successful policies without too great a show of enthusiasm. The tacit agreement to this effect between the party leaders gives our politics a certain air of unreality, and many of those who seek for truth in the mouths of politicians turn with relief to Russia. The government of the Soviet Union not only admits but boasts that its policy is experimental. Many items in its early programme were failures, and some of these have been withdrawn. Others equally daring in their conception have proved successful. Hence the evolution of the new social order has been amazingly rapid. The Communist party has been in power for less than ten years, but it has contrived to evolve a fairly stable system combining some of the advantages of capitalism and socialism. No doubt the Russian people has proved an ideal subject for large-scale experiments. But the growing distrust of constitutional government in Europe suggests that there, too, the present generation

is more prepared to be experimented on than were its
fathers. And if we are to escape the despotism which
will follow a revolution either to the Left or the Right,
our present rulers and those who support them will be
well advised explicitly to imitate the extremely capable
Bolshevik leaders, and adopt an experimental method.

In the sphere of ethics the same principles must, I
believe, be applied. The circumstances postulated by
the older ethical codes have ceased to exist. In a more
primitive community our most obvious duty was quite
literally to our neighbour. In a village we knew our
neighbour's affairs pretty well, and if we did not always
succeed in loving him as ourself we could pretty often
be of assistance to him. In a great city one may have
a department-store on the left and a man one never
meets on the right. An occasional gift to charity or
even an evening a week spent on welfare work in a
poorer quarter is not the psychological equivalent of
taking in Mrs. Johnson's children during her illness and
going to the assistance of Mrs. Kelly when her husband
comes home drunk. All through the civilized world
experiments are being made as to how best to help one's
fellow-creatures without falling into hard officialism on
the one hand or indiscriminate gifts to the undeserving
on the other. The mere multiplicity of these experiments goes to show how few of them have been completely successful.

Again, the invention of contraceptive methods and
the economic emancipation of women have created new
problems in sexual morality. If a given action has
different consequences now from those which would
have followed fifty years ago, it is from the ethical point

of view a different action. Contraception is leading to experiments on rather a large scale in Europe ; and most of them, like most laboratory experiments, are un- · successful. Married women are discovering that no children or a single child seldom leads to happiness ; unmarried women who try the experiment rarely find satisfaction in a multitude of lovers. On the other hand, a spacing out of child-births is generally found to be advantageous for all concerned, and there is a small but perhaps a growing body of experience favouring an experimental honeymoon before marriage in lands where divorce is difficult, and an experimental period of marriage where it can easily be dissolved. The public discussion of such topics generally leads to the promulgation by both sides of dogmatically held opinions and a failure to realize that the questions at issue can only be decided by experience. This failure is unfortunate for two reasons. It means that many more experiments in behaviour, often of a disastrous type, will be needed before the question is cleared up, than would be the case if a serious attempt were being made to collate the results of those going on to-day. And it is extraordinarily difficult to love one's neighbour when he or she differs from one fundamentally on moral issues, though quite possible to do so if one believes that he or she has made an unfortunate mistake in conduct because of uncertainty as to what, under the new conditions, was right or wrong.

Such then is the case, or rather a fragment of the case, for doubt. It is very nearly the same as the case for freedom of speech. Plato described thought as the dialogue of the soul with itself, and doubt is just a refusal to deprive either side of a hearing. Just as freedom of speech facilitates right action by the State, provided the speakers and those who listen to them have a share in deciding policy, so doubt is a virtue if, and only if, it is the prelude to action. A merely negative doubt is like freedom of speech divorced from political responsibility. This was the condition of affairs in India in the ten years before 1919, when the Indian politicians were permitted to talk indefinitely, but possessed no effective share in the government. India is barely beginning to recover from the type of political thinking which flourished during that unfortunate epoch.

There are some who will admit that doubt may be a necessity in a scientific era, but hold that art and literature flourish best in an age of faith when they become the interpreters of a great religious or philosophical system rather than the symptoms of intellectual unrest. While such opponents bring forward Dante and the architects of the European cathedrals, forgetting Milton and Phidias, I shall do no more than cite the opinion of John Keats in a letter to his brother : ' Dilke is a man who cannot feel he has a personal identity unless he has made up his mind about everything. The only means of strengthening one's intellect is to make up one's mind about nothing—to let the mind be a thoroughfare for all thoughts, not a select party.' Keats certainly did not strengthen his intellect at the expense of his aesthetic powers, and his *Hyperion* is little more than an account of the supersession of good ideas by better, a process which, as he showed, so far from stifling art, may inspire it.

Finally, I shall perhaps be told that I am preaching pragmatism. But where the pragmatist says that a belief is true because it works, I have attempted to suggest that it is often false although it works, and that belief is not, as James preached, a necessary preliminary to effective action. And where the pragmatist exalts the will to believe, I have attacked it. The desire for intellectual certitude is laudable in the young, as a stimulus to thought and learning ; in the adult it easily becomes a vice. History, when it is taught as the history of human thought, makes it abundantly clear that most of the intellectual certitudes of our forefathers were illusory, though often of temporary value. One intellectual certitude has from time to time been replaced by another at the expense of a sufficient number of martyrs. So long as our education aims as inculcating dogmas, religious, political, ethical, or scientific, fresh relays of martyrs will be necessary for every step of human progress. And while I do not suggest that humanity will ever be able to dispense with its martyrs, I cannot avoid the suspicion that with a little more thought and a little less belief their number may be substantially reduced.

To sum up, science has owed its wonderful progress very largely to the habit of doubting all theories, even those on which one's action is founded. The motto of the Royal Society, ' Nullius in verba,' which may be paraphrased ' We take nobody's word for it,' is a sound rule in the other departments of life. The example of science shows that it is no check on action. Its general adoption would immeasurably hasten human progress.

SCIENCE AND THEOLOGY AS ART FORMS

RELIGION and science are human activities with both practical and theoretical sides. There is at present a certain degree of conflict between them, and this will undoubtedly continue for some generations. During this conflict the disputants have tended to emphasize the differences between them. But their resemblances are equally interesting, and perhaps throw a good deal of light on the differences.

It is only very recently that they have had a chance of diverging. Readers of the Pentateuch, or of the contemporaneous or earlier religious literature of Egypt or Mesopotamia, will find it very difficult to disentangle the science from the religion. The Pentateuch contains some very good applied science in the sanitary laws of Moses. The palaeontology of Genesis is also correct in many points ; particularly in describing a period of the earth's history before the origin of life, followed by the appearance of animals in the seas, and only later on land, man being the last creation. It is of course wrong in putting the origin of plants before that of stars, of birds before that of creeping things, and in several other respects.

Unfortunately, however, since Moses' time science and religion have diverged, not without a certain loss to both. The reason for this divergence can best be seen when we study some typical scientific and religious minds at work. Each starts from a certain experience,

and builds up a system of thought to bring it into line
with the remainder of experience. The organic
chemist says, ' The substance I have just made is a
liquid with a characteristic smell, melting at 31° C.,
boiling at 162° C., and whose compound with phenyl-
hydrazine melts at 97° C. I have probably synthesized
furfural in a new way.' The saint says, ' I have had
an experience very wonderful and rather difficult to
describe in detail, but I interpret it to mean that God
desires me to devote myself to preaching rather than
to shut myself up from the world.'

The scientific man then starts from experiences in
themselves emotionally flat, though to him perhaps
interesting enough. He may end by producing a
theory as exciting as Darwinism, or a practical invention as important as antiseptics or high explosives.
The mind of the religious man on the other hand works
on a descending scale of emotions. The dogma,
prophecy, or good works which he may produce are
inevitably less thrilling than his religious experience.

It is more interesting to most minds to read or speculate about the distances of the stars than to measure the
positions of their images on a photographic plate. But
it is less interesting to read a work on justification by
faith or on St. Thomas's theory of transubstantiation
than to take part in a well-conducted church service.

Now, the rather dull raw material of scientific thought
consists of facts which can be verified with sufficient
patience and skill. The theories to which they have
given rise are far less certain. They change from
generation to generation, even from year to year. And
the religious opponents of science tend to scoff at this

perpetual change. In scientific thought we adopt the simplest theory which will explain all the facts under consideration and enable us to predict new facts of the same kind. The catch in this criterion lies in the word ' simplest.' It is really an aesthetic canon such as we find implicit in our criticisms of poetry or painting. The layman finds such a law as $\dfrac{\partial x}{\partial t} = \kappa \dfrac{\partial^2 x}{\partial y^2}$ much less simple than ' it oozes,' of which it is the mathematical statement. The physicist reverses this judgment, and his statement is certainly the more fruitful of the two, so far as prediction is concerned. It is, however, a statement about something very unfamiliar to the plain man, namely, the rate of change of a rate of change. Now, scientific aesthetic prefers simple but precise statements about unfamiliar things to vaguer statements about well-known things. And this preference is justified by practical success. It is more satisfactory scientifically to say : ' The blood-vessels in John Smith's skin are dilated by a soluble toxin produced by a haemolytic streptococcus growing in his pharynx,' than to say that he has scarlet fever. It suggests methods of curing and preventing the disease. But only a few people have seen a streptococcus, and no one has seen a toxin in the pure state. In physics, the most developed of the sciences, things have gone so far that many physicists frankly say that they are describing atom models and not atoms. Atoms themselves have the same sort of reality as chairs and tables, because single atoms can be seen if going fast enough. But when we come to their internal structure we can only say that they behave, in some important respects, as if electrons were going round in them with such and such velocities in such and such orbits. If that is the real structure we can calculate the velocities with a great deal more accuracy than that with which a speedometer gives the speed of an automobile ; and verifiable predictions based on these calculated speeds come out with very great accuracy. But the speeds are not observable, and physicists are becoming less and less careful as to what hypothetical events they postulate to explain observable phenomena, provided the hypotheses enable them to predict accurately.

Einstein showed that we could explain and predict slightly better if he substituted other conceptions for those of space and time, and his own substitutes will doubtless be replaced in their turn. Now, if Einstein is right, or even partly right, no physicists before his time knew quite what they were talking about when they used the ideas of distance and time, and practically every statement that they made which purported to be accurate was false. So presumably is every such statement of a modern physicist. Similarly, chemists supposed that the weight of a chlorine atom was 35·46 units until Aston showed that chlorine atoms were a mixture of two kinds whose weights are 35 and 37. Almost all the deductions from the premise were right, but the premise itself was wrong. I have no doubt that biological theory is equally riddled with falsehoods.

In fact, the experience of the past makes it clear that many of our most cherished scientific theories contain so much falsehood as to deserve the title of myths. Their claims to belief are that they contradict fewer known facts than their predecessors, and that they are of practical use. But there is one very significant feature of the most fully developed scientific theories. They tell us nothing whatever about the inner nature of the units with which they deal. Electrons may be spiritually inert, they may be something like sensations, they may be good spirits or evil spirits. The physicist, however, can only tell us that they repel one another according to a certain law, are attracted by positive charges according to another law, and so on. He can say nothing about their real being, and knows that he cannot.

It is, I suppose, the fact that there is no great stability to be found in scientific theories which leads the opponents of science to talk of the intellectual bankruptcy of our age; generally as a preliminary to adopting beliefs current in mediaeval Europe, India, or Bedlam. We must therefore proceed to examine the claims of the theological beliefs which are offered as substitutes or supplements for science. Religious and moral experience are facts. Most people can obtain a certain amount of religious experience by a very moderate effort. It would be ridiculous not to interpret it. But its interpretation is still in a pre-scientific stage. There is something true in theology, because it leads to right action in some cases, and serves to explain certain otherwise difficultly explicable facts about the human soul. There is something untrue, because it often leads to wrong actions and leaves a great deal which is within its province, for example the origin of evil, unexplained.

In India and in pre-Christian Europe theology developed gradually. Gods or devils could be postulated as required. This, however, led to a polytheism incompatible either with the unity of nature or the unity of duty. Since about A.D. 400, however, Christian theology has altered very little. But the moral consciousness of Christians has altered a great deal. For example, it now condemns slavery and cruelty to animals. St. Paul condoned the one and ignored the other. But it is alleged by theologians that our moral consciousness is one source of our knowledge of God. If the one develops, so should the other. Now, in the early or growing stages of a religion a large number of myths are produced which interpret the moral and religious consciousness of the time. Jesus habitually used vivid imagery. Sometimes he was obviously speaking in parables. Sometimes a doubt exists, as with regard to the words used at the Last Supper. If taken as a statement of fact, they lead to a belief in transubstantiation. If not, they are merely an expression of solidarity like ' I am the vine, ye are the branches.' Some of his other statements are taken literally by all Christian churches, and it is claimed that a belief in them is a more or less essential pre-requisite to a Christian life. This appears to be doubtful.

But where Jesus used parables as far as possible, his followers stated innumerable doubtful propositions as facts, and attached more importance to a belief in them than the followers of the non-Christian religions have attached to similar statements. Some of these propositions were about the structure of the universe, e.g. ' The Holy Ghost proceeds from the Father and the Son,' others about events, e.g. ' Jesus descended into Hell.' Scientific men clearly cannot complain of the theologians for making myths. They do it themselves. Sometimes they take them very seriously. Occasionally the public does so. The ether is a highly mythical substance filling empty space, for which relativistic mechanics have little use. But broadcasting has made it a popular myth. To-day probably more men in England believe in it than in Jesus' bodily ascension to heaven.

The main objection to religious myths is that, once made, they are so difficult to destroy. Chemistry is not haunted by the phlogiston theory as Christianity is haunted by the theory of a God with a craving for bloody sacrifices. But it is also a fact that while serious attempts are constantly being made to verify scientific myths, religious myths, at least under Christianity and Islam, have become matters of faith which it is more or less impious to doubt, and which we must not attempt to verify by empirical means. Chemists believe that when a chemical reaction occurs, the weight of the reactants is unchanged. If this is not very nearly true, most of chemical theory is nonsense. But experiments are constantly being made to disprove it. It obviously cannot be proved, for, however accurately we weigh, the error may still be too small for us to observe. Chemists welcome such experiments and do not regard them as impious or even futile.

Christians almost all believe that prayer is sometimes answered. Now, all prayers are not answered, and desired events often occur for which no one has prayed. Hence individual instances of answered prayers are useless. We must have statistical evidence. It has been proposed from time to time that a group of believers should pray for the recovery of the patients in one wing of a hospital over a period of some months, and the number of deaths in it be compared with that in the other wing. The experiment has always been refused, partly on the ground that 'Thou shalt not tempt the Lord thy God,' partly through lack of faith. Until it has been made, I do not propose to ask for the prayers of any congregation on my behalf.

In the absence of experimental evidence Galton attempted a statistical investigation. He considered that of all classes of society in England those most prayed for were the sovereigns and the children of the clergy. If prayer is effective they should live appreciably longer than other persons exposed to similar risks of death. So kings were compared with lords, and the children of the clergy with those of other professional men. The conclusion to which his numbers led was that these much-prayed-for persons had slightly shorter lives than those with whom he compared them. The difference was not, however, great enough to make it probable that prayers have any harmful effect.

On the other hand, there can be little doubt that either one's own prayers or the knowledge that others are praying for one may serve as a safeguard against temptation, or effect the complete or partial cure of functional diseases of the nervous system. There is therefore, at any rate, a certain truth in the efficacy of prayer; but this efficacy may be explained on psychological grounds, and does not necessarily imply a divine interference in the order of nature.

Nevertheless, religious experience is a reality. It cannot be communicated directly, but those who experience it can induce it in others by myth and ritual. This fact is most fully recognized by Hindus. The simpler of them believe in a vast and complex system of myths, and attempt to gain their personal ends by placating various deities. The more intellectual do not pretend that the myths are true, or that the ritual has any effect other than a psychological influence on the participants. A well-known Hindu mathematician, who would sooner have died than eaten beef, once asked a British colleague whether it would cause real surprise in England if the Archbishop of Canterbury were to deny the existence of God in the House of Lords. In India, it appears, the corresponding event would excite no more comment than does a denial of the existence of the soul by a professor of psychology, or of the reality of matter by a professor of physics, in England or America.

This essay is emphatically not a defence of Hinduism, which has excused every kind of evil, from murder and the prostitution of young girls to self-mutilation and refusal to wash. It is, moreover, the chief prop of a system of hereditary class distinctions based in part on differences of colour; by contrast with which Louis XIV. appears an equalitarian and Pizarro a champion of racial equality. And I find many of its myths disgusting. But the fact remains that it has lasted a great deal longer than Christianity and shows far fewer signs of decay. It has more than half as many adherents, and on the whole affects their lives to a greater extent.

One of the gravest errors into which the Christian religion has fallen is the view that a myth or dogma cannot influence you unless you believe it. Christians in general believe in the existence of St. Bartholomew and St. James the Less, but they are much less influential people at the present day than Cordelia and Father Christmas, in whom very few adults believe. If the Christian churches continue to make belief the test of religion, it appears to me that one of three things will happen.

If they maintain their influence they will sterilize scientific thought, and either slow down human progress or render their adherents as defenceless against non-Christian races armed with science as were the Asiatics against the Christian races during the nineteenth century. The Caliph Mutawakkil, who, perhaps more than any other man, assured the triumph of orthodoxy and the suppression of independent thought in the Mohammedan world, was responsible for its conquest by the Christians, when these latter were allowed to think for themselves about the nature of matter and hence to produce steam engines and high explosives. The Christian churches are preventing people from thinking for themselves about life. In the interests of theological orthodoxy Christian children may not be taught about evolution. In those of morality and decency they may not be taught how their bodies work. If any Asiatic people begins as a whole to think biologically before those of European origin do so, it will dominate the world, if the lessons of the past are any guide to the future.

It is perhaps more probable that under such circumstances religion will decline and become unimportant in human life. This would, perhaps, be a misfortune, for it is probably the essence of religion that a man should realize that his own happiness and that of his neighbours are not his only concerns. If we abandon religion completely we shall probably deify ourselves and behave in a wholly selfish manner, or deify the State and behave as the more violently patriotic Germans did in the late war. (And as the war continued the other nations conformed more and more to the German model.)

It may be that religion as we know it will some day be superseded altogether, for many people nowadays lead good lives without it. It is possible to become convinced on philosophical grounds of the supreme reality and importance of the spiritual, without postulating another world or even a personal God. And such convictions may be supported by mystical experience. However, it is improbable that most people are capable of the abstraction necessary if such a point of view is to be adopted. It is even harder, though of course possible, to worship one's own moral convictions, as Mr. Bertrand Russell does, while believing that they are an unimportant by-product of a universe which, as a whole, is indifferent to them. One of these views of the world may well be true, but I do not believe that most people can attain to it for some centuries. And self-worship and State-worship are within the reach of all.

A third possibility is the rise on the ruins of Christianity of a religion with a creed in harmony with modern thought, or more probably the thought of a generation ago. Traces of such a creed may be found in the utterances of prominent spiritualists, in the economic dogmas of the communist party, in the writings of believers in creative evolution, and elsewhere. A new religion would crystallize the scientific theories of its own age. The old religions are full of outworn science, including the astronomical theory of a solid heaven, the chemical theory that water, bread, books, and other objects can be rendered holy by special processes, and the physiological theory that a substance called a soul leaves the body at the moment of death. I remember hearing the headmaster of Eton informing the school from the pulpit that the body lost weight (or gained it, I forget which) at the moment of death. This materialistic view of the nature of the soul is quite prevalent in Christian circles. A new religion would probably include in its creed the reality of the ether, the habitability of Mars, the duty of man to co-operate in the process of evolution, the existence of innate psychological difference between the human races, and the wickedness of either capitalism or socialism.

Now, Christianity has ceased to obstruct astronomy and geology with such texts as ' He hath established the world so fast that it cannot be moved,' and ' The waters under the earth.' It will find it a harder task to admit that it has been as wrong about the nature of the soul as about the nature of heaven. Nevertheless, such an admission will have to be made if Christianity is to survive as the religion of any appreciable fraction of educated men and women. The experience of the past shows that such an admission is not impossible. In the words of Renan, ' Si le parti radical, parmi nous, était moins étranger à l'histoire religieuse, il saurait que les religions sont des femmes dont il est très facile de tout obtenir, si on sait les prendre, impossible de rien obtenir, si on veut procéder à haute lutte.'

On the other hand, the dogmas of a new religion would include enough of contemporary science to be fairly plausible, and enough of contemporary ethics to be fairly practicable.   It would thus constitute a far more serious obstacle than Christianity to scientific and ethical progress in the future, even if it led to a momentary advance in scientific and ethical education.

I therefore suggest the following standpoint for con-sideration. Religious experience is a reality. Hence it affords a certain insight into the nature of the universe, and must be considered in any account of it.   But any account given of religious experience must be regarded with the gravest suspicion. Some of the greatest mystics described their experience in language which is frankly and admittedly self-contradictory.   Such ex-pressions as ' a delicious desert,' ' a dazzling darkness,' ' ein lauter Nichts,' ' it neither moves nor rests,' are common in their writings.   Others have admitted their ignorance of details. ' Nescio, nescio, quae jubilatio, lux tibi qualis ' (' I know not, I know not, what is thy joy and thy light '), said Bernard of Cluny.   Where a more definite statement is made, it is even less reliable than the crude interpretation of our sense data.   Our ordinary perceptions tell us that the sun, a body of in-determinate size, but certainly not as large as a country, let alone the whole world, rises every morning from a level below our own and descends again at evening.   I suggest that the data of religious experience as to the nature of God and his relation to man are no more reliable than those of crude perception concerning the sun and its relation to the earth.   In fact, they are less so, for different mystics disagree with one another. Nevertheless, mystical experience may be as capable of scientific investigation and explanation as sensuous experience, though the process would be more difficult because the experience is rarer. Such an interpretation, provided its results did not contradict those of other branches of knowledge, would constitute for the first time a theology worthy of the name of a science. Whether its conclusion would be theistic or not it is impossible to judge.

The theology, or rather theologies, of to-day are something quite different. Religious experience has so far been described mainly in symbols drawn from every-day life. Some such symbolic expression seems to be inevitable. Unfortunately, the vast majority of mystics, or at least their disciples, have come to take the symbols seriously. Some modern religious literature, however, furnishes a most gratifying exception to this rule. The writings of L. P. Jacks, for example, which undoubtedly sprung from a considerable body of religious experience, are mainly cast in the form of fiction. His account of a future life in *All Men are Ghosts* is the most attractive that I have so far come across, but it does not purport to be accurate. The similar accounts given by the saints are also mostly, if not all, based on genuine religious experience, but they are not for that reason to be believed. Theology or mythology is an art form used to express religious experience. Both theology and scientific theory may be valuable guides to conduct, as well as beautiful in themselves, but that does not make them true. The theory that heat is a substance, not a mode of motion, enabled Watt to make the calculations on which the design of his steam-engines was based. The calculations were right, but the theory was wrong. The Christian saints believed God to be a person (though the Buddhists did not). Their lives were often, though not always, admirable ; and their religious experience on the whole conformed with their beliefs. But it does not follow that their beliefs were correct.

If such a point of view is adopted, the literature of Christianity will come to be regarded as of mainly symbolical value ; but yet as showing forth a real experience which could perhaps have been expressed in no other way, at the time when it was composed. An ever larger proportion of the sayings of Jesus will be regarded as parables and therefore to be interpreted ; rather than dogmas to be believed, and then fitted, not without strain, to the rest of experience. Christians will learn to take many of the churches' prohibitions no more seriously than St. Paul's veto on things strangled. They will regard their institutions rather as tokens of solidarity with the past and the future than as means of salvation. And they will rank theology with poetry, music, rituals, architecture, sculpture, and painting, as an expression of religion, but not its essence.

Perhaps a summary of the ideal relationship of religion and science would be somewhat as follows :— Religion is a way of life and an attitude to the universe. It brings men into closer touch with the inner nature of reality. Statements of fact made in its name are untrue in detail, but often contain some truth at their core. Science is also a way of life and an attitude to the universe. It is concerned with everything but the nature of reality. Statements of fact made in its name are generally right in detail, but can only reveal the form, and not the real nature, of existence. The wise man regulates his conduct by the theories both of religion and science. But he regards these theories not as statements of ultimate fact but as art forms.

MEROZ

'Curse ye Meroz,' said the angel of the Lord; 'curse ye bitterly the inhabitants thereof; because they came not to the help of the Lord, to the help of the Lord against the mighty.'—JUDGES v. 23.

IN recent years defenders of organized religion have shifted their ground. Noah is relegated to the children's section of the *Daily News*. The resurrection of the body is believed in in a Pickwickian sense. But we are told that the churches, or one of them, witness to the presence on earth of the Holy Spirit, and that when every allowance has been made for antiquated dogma they still represent something worthy of our adoration. If we cannot go so far as that, we are asked at least not to oppose their activities, on the ground that they do more good than harm. That they do some good is, of course, undeniable, and for a number of years this latter argument carried weight in my own case. I will try to explain why it has ceased to do so.

I should be the last to suggest that the late war was a good thing, but there is no doubt that it furnished a rough test of character. It will therefore be interesting to analyse the conduct of ministers of religion during its course. Their theological views may be incorrect, but if they conduce to nobility of character they are excusable. ' By their fruits ye shall know them.' In this analysis religious bodies fall fairly sharply into three groups, the national churches, the pacifist churches, and the Roman Catholic Church. The former include not only officially recognized bodies such as the Orthodox Church of Russia and the Church of England, but those whose conduct was substantially similar ; for example, most of the British nonconformist bodies. Their priests and ministers as a whole enthusiastically backed their respective countries. I do not blame them for that. They were subject to the same emotions as the rest of us. But whereas those emotions led large numbers of laymen to death and mutilation, their effects on the alleged servants of Jesus were generally less drastic. I can only speak of the clergy of the British Empire with any first-hand knowledge. Things were, I gather, much the same in Germany and the United States of America. In Servia, for example, they may well have been different.

A large number of the younger clergy became army chaplains. In this way they at once obtained the very satisfactory status of commissioned officers. With other officers that status was on the whole a fair return for the very grave dangers which they ran. The army chaplains generally ran the irreducible minimum of risk. Most of them kept very well behind the line. In my war experience I never saw a chaplain display courage. I once saw a doctor display cowardice, but this was after he had been shelled for twelve hours on end, and then blown up. There were, of course, exceptions among the clergy ; I am proud to number two of them among my acquaintance. The standard of courage was, I believe, much higher among the Catholics than among the Protestants. The former repeatedly heard confessions under machine-gun fire in the open.

The latter very rarely took the opportunities offered them to promote the cause of religion by risking their lives. For one dead padre created a great deal more impression than a dozen living. I am quite aware that several chaplains won the V.C. These men are typical of the thoroughly sincere and idealistic men who still enter the Church, though in diminishing numbers. They were no more representative of their profession as a whole than the chaplain who was found rifling the pockets of British and Bulgarian dead at Yenikeui in September 1916.

It may be contended that I was unfortunate in my acquaintance among army chaplains, and biassed in my interpretation of their conduct. It is therefore important to examine the behaviour of the clergy as a body. When conscription was introduced in Britain the clergy of all denominations showed a unanimity without parallel since the Reformation. Conscription was not for them, and so great was clerical influence among the governing classes that their exemption was taken as a matter of course. Now, from a Christian point of view, it is perhaps arguable that ministers of religion should not fight. But there is absolutely no reason why the self-styled disciples of Jesus should not, as privates of the R.A.M.C., have tended the sick and wounded under conditions of moderate discomfort if relatively little danger.

The attitude of a person who stays at home while encouraging others to go to war is reminiscent to the lay mind of that of the Duke of Plaza Toro; but it might arouse less disgust if it were assumed, not merely during a popular war, but during an unpopular one, provided it were equally justifiable from the Christian point of view. The test came in 1922, when a strange event occurred. Mr. Lloyd George took an honourable but inexpedient course. He proposed, as he was in honour bound, to support the defeated Greeks against the Turks. His Conservative colleagues, who, with a smile or a sigh, had forfeited any title to be called Unionists during the preceding three years, turned upon him and brought him down. A war against Turkey would not have paid. But we were as much bound by honour to support Greece in 1922 as France in 1914, though our case for supporting Belgium was no doubt better. And this support would probably have saved hundreds of thousands of Christians from death and rape, and certainly have rendered Eastern Thrace a Christian instead of a Mohammedan country. Two or three clergymen above the military age, including the present headmaster of Eton, took the view that such considerations possessed a certain importance. The remainder appear to have convinced themselves that they could serve both God and Lord Rothermere. An ' Anglo-Catholic ' acquaintance informed me at the time that he attended several requiem masses for the souls of the Christians killed at Smyrna. But it has been the usual Christian custom to pray to, and not for, martyrs. And since the process has been reversed, I have ceased to attend the services of a church which no longer takes itself seriously. As I do not prefer the Eastern form of Christianity to Islam, I am personally very glad that we did not fight the Turks.

It is interesting to compare the clerical attitude to the Turks with their reaction to the Bolsheviks. The former have killed many more Christians and many more clergy than the latter, and have successfully abolished Christianity over large areas. It cannot even be claimed that the Turks, being good Muslims, are preferable to the infidel Bolsheviks. The present rulers of Turkey do not conform to most of the precepts of Islam, and Mustapha Kemal's treatment of the Caliph, and of the revenues of Mohammedan institutions, would have gladdened the heart of Peter the Hermit. But the Turks have not adopted Communism, and our clergy have less personal danger to fear from the followers of Kemal than from those of Lenin.

The pacifist churches have a better record. The Society of Friends is the only religious body which has come out of the war with an enhanced reputation. A pacifism which often took the form of mine-sweeping, or driving ambulances under fire, earned the respect of the soldiers. But the Quakers are almost alone among Christian churches in possessing no clergy. A few obscure American sects adopted a pacifist attitude on the ground that the world was about to end, and the battle of Armageddon would be more satisfactory for lookers-on than for participants. They were ruthlessly and successfully persecuted, and their leaders imprisoned. The survival of our planet removes their claims to consideration.

We finally come to the Roman Catholic Church, for which, as a political institution, I feel the greatest admiration. The authorities of the Church had to fear a Russian victory and the break-up of Austria. The latter event has lost them millions of adherents in Czecho-Slovakia ; the former would have lost them millions in Poland, which has celebrated its independence by rooting out the Orthodox Church. (It was the Catholic Poles, not Lenin or Calles, who demolished the Greek Orthodox Cathedral at Warsaw without any serious protest from other Christians.) On the other hand, open support of the Central Powers would have alienated many of the Catholics opposed to them. The Church's problem was therefore how to remain officially neutral while giving the maximum support to its allies. A few facts came out from time to time which might have shaken a faith that was not supernatural. The Vatican was in close touch with Austria in July 1914, and on July 26th the Bavarian Minister at the Vatican telegraphed to Munich that the Pope approved of energetic action against Servia.

During the war Mgr. Gerlach, the Papal Master of the Robes, used his sheltered position in the Vatican to plan the blowing up of two Italian battleships. In Ireland, Canada, and Australia the Catholic clergy organized the resistance of conscription with varying degrees of success. They did not do so in Alsace, nor in German and Austrian Poland, and at the present day the anti-French movement in Alsace is largely organized by the Catholic clergy. It is quite natural that this should be the case, since the French treat them worse than did the Germans, on the simple ground that, in the words of Pope Leo XIII., ' If the laws of the State . . . violate the authority of Jesus Christ, vested in the Sovereign Pontiff, it is a positive duty to resist, a crime to obey.' And the French, who have made their own laws for the last fifty-six years, wish them to be obeyed.

Some hope might have been placed in Judaism, the

other great international religion. A world-wide
financial operation in July 1914 might have prevented
or at least hampered the outbreak of war. But wherever they were allowed to do so, that is everywhere
outside Russia, the Jews showed as much patriotism as
any one else, though perhaps they wearied a little
sooner. And the British rabbis shared all the immunities of their Christian colleagues.

In this crucial test, then, the only religious organization which emerged with credit was the priestless
Society of Friends, and the clergy consistently displayed
a far lower standard of morality than their flocks.
There is nothing surprising in this. The Catholic
(and Anglo-Catholic) clergy claim a spiritual descent
from the Jewish priesthood, whose best-known members
were Annas and Caiaphas.

Priests have always used their power to evade the
moral obligations of the ordinary man ; and threatened
him with fire here or hereafter, or with social or economic penalties, if he referred to the fact. What is
new in the situation is that the public is beginning to
recognize the moral and intellectual inferiority of the
clergy. Their income is diminishing, and it is not at
all likely to increase. For whereas the clergy of sixty
years and over are on the whole men of fair intelligence,
those of to-day are being recruited from the dregs of the
universities, whilst many have no higher education
at all. Under these circumstances they are hardly
likely to tap fresh sources of revenue.

If Protestant Christianity is to be saved, it will not be
by its clergy, but by men who, like St. Paul, will preach
the Gospel in the intervals of earning their living and

risking their lives like ordinary mortals. They would not be compelled by economic considerations to profess dogmas which become daily less credible. And they might salve what is valuable in Christianity from the present wreckage. But the longer the fortunes of that religion are entrusted to the clergy, the more remote does that contingency become.

SOME ENEMIES OF SCIENCE

LAST week my wife successfully poisoned a number of rats. They were eating the food of our chickens, and would have eaten the smaller of the birds if they had got the chance. Owing to the failure of a more humane poison she found herself compelled to use phosphorus ; which is a slow and, to judge from the experience of human beings who commit suicide by eating matches, often rather painful means of death.

During the same period I killed two rats in the course of experimental work intended to advance medical science. One of them, if we can judge from human experience (and we have no more direct means of evaluating the consciousness of animals), died after a period of rather pleasant delirium like that of alcoholic intoxication. The other had convulsions, and may have been in pain for three or four minutes. I should be very thankful if I knew that I should suffer no more than it did before my death. It therefore seems to me somewhat ridiculous that, whereas my wife is encouraged by the Government and the Press, I should be compelled to apply to the President of the Royal Society and another eminent man of science for signatures to an application to the already overworked Home Secretary, before I can even kill a mouse in a slightly novel manner.

It is probably right that some control should be kept

over experiments likely to involve severe and prolonged pain to animals; but it is monstrous that with regard to wholly or nearly painless procedures the scientific man should be worse treated than any other member of the community.

Under the present law, or, at any rate, under the law as at present interpreted, a licence is required for a large number of absolutely painless experiments, and, what is more serious, they can only be performed in a limited number of laboratories. In consequence, the isolated amateur worker, who has played so great a part in the development of British science, is debarred from wide fields of physiology. The sportsman may go out and shoot as many rabbits as he pleases ; and if some of them are wounded and escape to die a lingering death in their holes, no blame attaches to him. But if he anaesthetizes one of his own rabbits at home, and opens its abdomen to observe the effect of a drug on its intestines, killing it before it recovers consciousness, he will be lucky if he escapes with a fine.

Nor may the doctor, after his day's shooting of unanaesthetized partridges, acquire surgical skill by an operation on an anaesthetized animal, even in a licensed laboratory. He has to practise on human patients. There are, of course, a few operations of human surgery for which animals would furnish relatively little guidance. In the majority of cases, however, they would be of very real value, and have been proved to be so in America. Not only is medical science already greatly hampered by the law, but a constant fight has to be kept up to preserve what possibilities are left it.

It is worth while enquiring into the reasons which have led to this state of affairs.

There are a few honest anti-vivisectionists. They are, of course, vegetarians ; for the painless killing of animals for physiology is no more reprehensible than their killing for meat. They wear canvas shoes, cotton or woollen gloves, and artificial pearls if any. They refuse to sit on leather-covered chairs, or to wear horn-rimmed spectacles. They do not spray their roses, nor employ Keating's powder even under the gravest provocation. I have not met any of them, but I am quite prepared to believe that they exist. No one who does not come up to this rather exacting standard can logically demand the total abolition of vivisection. But logic is not the strongest point of the enemies of science.

All others who demand the prohibition of experiments on anaesthetized animals are quite definitely hypocrites, engaged in the familiar pursuit of

' Compounding sins they are inclined to
By damning those they have no mind to.'

There are few more disgusting spectacles in our public life than that of the two or three sporting peers who habitually introduce or support Bills to prohibit such experiments. Each of them has caused more pain to animals in a single day's sport than the average physiologist inflicts in a lifetime, and usually for no end except his personal pleasure. For it seems to me that from the ethical point of view a fairly sharp distinction can be drawn between the killing of animals bred for this purpose at a considerable expense, which would produce far more food if applied to agriculture ; and that of rabbits, hares, and pigeons which must be kept down in the interest of crops and livestock. Personally, since I have realized from my own experience with shell splinters that it is no fun to carry bits of metal about one's person, I would no more shoot a rabbit than kill my bacon for breakfast. But I certainly do not condemn those who do so.

We must next consider the relatively small number of anti-vivisectionists who would merely prohibit all painful or possibly painful experiments. Now, the world is so constituted that we cannot avoid inflicting pain on others. I cannot dig in my garden without bisecting a number of earthworms, or drive a car for any time without running over a few of the various animals whose flattened corpses decorate our country roads. But it is our duty, as far as possible, to diminish the amount of pain in the world. The question therefore is whether medical research does this or not.

Now, anti-vivisectionist literature distorts both sides of the account. It states that a great deal of severe suffering is inflicted on animals in the name of science, and that there is little or no return for this in the diminution of human and animal suffering. With regard to the first of these assertions I can speak with a certain degree of experience. I have seen numerous experiments on animals, but I have never seen an animal undergoing pain which I would not have been willing to undergo myself for the same object. Why, then, it may be asked, should not all painful experiments be done on human volunteers ?

There are several reasons why not. One is the very simple fact that many of these experiments possibly or necessarily involve the death of the animal. For example, rats are frequently inoculated under the skin of their sides with transplantable cancers. These are not painful, for the rat does not wince or squeak when the lump is pressed. If it were allowed to die of cancer it would often suffer; for the original tumour or its metastases elsewhere would press on nerves, and one of them would probably start to ulcerate. But before either of these events occurs, all such rats in the laboratory in which I work are killed, and the tumours used for chemical study or inoculation. A man, even if he could legally be used for such a purpose and chloroformed before pain began, would presumably suffer from the anticipation of an early death.

Just the same applies to deformity. A rickety child suffers mainly because it cannot take part in the activities of its comrades and is made to realize that it is deformed. A rickety rat has none of these disadvantages, not only because it is probably not self-conscious, but because under laboratory conditions all its acquaintances are rickety too. Finally, there is the question of expense. Human beings cost a lot in board and lodging, and must be compensated for loss of time. If, as in experiments on the effects of small changes in the diet, hundreds of individuals and years of time are needed, this consideration is generally final.

However, if we are to believe anti-vivisectionists, animals constantly undergo tortures which no human being would voluntarily endure. I recently received an illustrated pamphlet, which I should think is fairly typical, describing the sufferings of laboratory animals. There was a picture of an oven in which dogs were slowly heated till they died, while a physiologist watched their agonies through a window. The thought of such cruelty would have made my blood boil, if it had not already been partially boiled in such a chamber on several occasions. Under such circumstances one becomes dizzy long before there is any definite pain, and death, if it occurs, is from heat stroke, not from burning. Personally, I prefer being overheated in a bath. Immersion of all but the head in water kept hot enough just not to be painful, causes loss of consciousness, after a good deal of panting, in about twenty minutes. Hence there is reason to think that a lobster, if put into cold water and heated fairly slowly, feels no pain, which it must certainly do if dropped into boiling water. Probably, however, it would suffer still less if about $2\frac{1}{2}$ per cent. of salt were added to the cold water.

Then came a picture of a dog's mouth held open by a somewhat brutal-looking contrivance. This was said to be taken from a scientific periodical called the ' Transactions of the Physiological Society.' There is, unfortunately, no such journal, nor could I find the picture in the Proceedings of that body for the date given. Perhaps, therefore, it was the anti-vivisectionist's idea of what an instrument of scientific torture ought to look like. But if I had been a maiden lady with a pet dog and no knowledge either of the facts or the literature of physiology, I might have sent a cheque to one of the ladies and gentlemen who make a living by compiling documents of this kind.

In some cases experiments are supposed to be painful out of ignorance rather than malice. A group of experiments by Sir John Bradford, in which parts of the kidneys of dogs were removed under an anaesthetic, are constantly described in Parliament as torture. Some of these dogs recovered completely, others died with the symptoms of chronic kidney disease, which in human beings seldom causes any pain worse than a headache. Stone in the kidney can, of course, be very painful, but the dogs were not so treated as to cause them pain of this type, nor did they show any signs of suffering it. As a matter of fact, too, dogs can stand a good deal of wounding without much suffering, so far as one can judge. I know this, not from laboratory experience, but because I have owned a dog whose courage and love affairs constantly led him into fights with larger dogs.

A large part of the unhappiness of dogs in English laboratories is directly due to the anti-vivisectionists. In the laboratory where I work there are a number of dogs, each of which, for two or three months in the year, eats certain organic compounds which it transforms in its body. The newly formed compounds are excreted in the urine. To facilitate the collection of urine an operation has been performed on them analogous to circumcision, and not nearly so severe as tail-docking. Of course, an anaesthetic was used. But because the operation has been performed in the cause of science rather than fashion, these dogs are forbidden by law to leave the laboratory. They are exercised in the grounds twice daily, but may not go into the street, and must lead a rather dull life. This regulation is typical of the present law, which is designed quite as much to hamper research as to protect animals.

While the large majority of experiments performed

annually are nearly painless, a few dozen, which attempt to reproduce a painful human disease, and thus to discover its cause or cure, are as painful as the disease which they imitate, except that when the animal's condition is clearly hopeless it can be killed.

I do not think it will be necessary to convince any reader of this book of the value of medical research. It has been the principal cause which renders the worst slum of to-day healthier than the palace of a century ago. If that result had been reached by the infliction of appalling torture on millions of animals the ethical justification of this torture would certainly be a matter for discussion. Actually the fate of experimental rats, for example, is no worse than that of pet rats, which generally die from deficient diet or epidemic disease.

It remains to consider the psychology of anti-vivisectionists. I think that their most important motive is a hatred of science, which they attack at its weakest point. They hate science partly because they do not understand it, and will not take the trouble to ; partly because it is ethically neutral. Many of them feel that disease must be a punishment for sin, and that it could be avoided if we lived according to their own particular prejudices. This view has been taken by most religions, though, of course, Jesus did not share it (John, chap. 9, v. 3). Almost all believe that there is some short cut to health. So a great many simple-lifers, vegetarians, faith-healers, Christian scientists, and so forth, are opposed to medical research, and say that its results are worthless.

In some cases anti-vivisection goes with pacifism. The fallacy involved in this association is rather interesting. Non-resistance of human evil is sometimes effective. A certain percentage of human smiters are seriously disconcerted if one turns the other cheek to them. But this kind of method does not work on bacteria, which have no finer feelings. We cannot find out how they behave, and thus acquire power over them, except by experiments on men or animals. This is a very unfortunate fact, but then the universe differs in a great many ways from what we should wish it to be. Medicine continued on non-experimental lines (with a very few exceptions) from the dawn of history till the seventeenth century. And in consequence it remained stationary during thousands of years. If its enemies get their way it will begin to stagnate again.

But there is a less respectable side to the anti-vivisectionist mind. During the recent agitation against experiments on dogs I made an offer of £100 (published in the *Daily Mail*) to the National Canine Defence League, if they would produce any evidence for certain libels on the medical profession which they were circulating in order to obtain signatures for a petition. I got no answer from the League, but a number of abusive and most instructive letters.

One of them, from E. Hough, of Hammersmith, objected to experiments ' on the dear, faithful, doggies for the benefit of worthless human beings.' ' I like to think,' she wrote (for I picture the writer as an elderly and soured spinster) ' that God will torture physiologists in a future life. I would not lift a finger to save one of them if he were writhing in agony.' There is, then, a group of anti-vivisectionists who like to think about torture. As they can no longer attend the burning of atheists and witches, they gloat over imaginary stories of animal torture till their blood boils ; and then cool it with the thought of physiologists in hell fire. Thanks largely to the psychological mutilation to which our society subjects adults, and more particularly children, the world is over-full of

'Ceux dont le rêve obscur salit tout ce qu'il touche,'

and I suspect that a fair number of them become anti-vivisectionists.

Those who have benefited by the results of medical research and wish it to continue might do worse than support the Research Defence Society, which carries on a lonely fight against a vast flood of lies. And they should urge the following alterations in the law, none of which would increase animal suffering in the faintest degree. Stray dogs impounded by the police should be used for experiment. This would abolish dog-stealing for laboratories, and save the lives of some thousand dogs per year. No licence should be required for experiments on fully anaesthetized animals which are killed under the anaesthetic. Surgeons should be allowed to practise their art under the same conditions. Animals should not be condemned to imprisonment for life because an experiment has been done on them. And in the interests of national economy the number of officials and of Government forms used in the supervision of research should be cut down.

At present biological and medical research workers are enormously handicapped by the law and by public opinion. Several hospitals, out of deference to subscribers, do not allow animal experiments. They thus render the rapid diagnosis of various diseases impossible, and kill a certain number of patients annually. And medical teaching is seriously handicapped in the same way. These are some of the reasons why England is less healthy than a number of other European countries. Anti-vivisectionists are responsible for far more deaths per year in England than motor vehicles, smallpox, or typhoid fever.

POSSIBLE WORLDS

IT is not clear that professionalism is any more desirable in philosophy than in football or religion. The professional philosopher tends to use mental processes of a type which has proved rather a failure in scientific thought. If I want to find out how my body works I alter it in some well-defined physical or chemical way, *e.g.* heat it through three degrees, or drink two gallons of water. If the mathematician doubts the validity of an argument which proves the convergence of an infinite series satisfying a given criterion, he constructs a series which obeys the criterion but does not converge. Such tests are conclusive, and have shown up the inaccuracy of some trains of reasoning which were at first sight very convincing. The same method can be applied in metaphysics. Renan, in the preface to his *Dialogues Philosophiques*, said that he intended to write a book called ' Hypothèses ' in which seven or eight world systems would be sketched, each lacking an essential principle. In this way he would demonstrate the importance of the missing principle more clearly than was possible by mere argument.

Unfortunately, he never carried out his admirable project. A certain amount has been done in the way of parables and myths, but posterity insists on taking them seriously. Witness the story of Dives and Lazarus, excellent as a parable, in which form it was delivered, but lamentable when taken seriously, as it has been

260

during nearly nineteen centuries, as a concrete picture of the future life. Generally, philosophers who construct a funny world come to believe that it is the real world. They find few to agree with them, and it is unfortunate that the whole life of a philosopher should be devoted to a single intellectual experiment. Even a rabbit can often be used for several, provided it is not irreversibly damaged. But philosophers damage their minds by coming to believe in their own hypotheses. This is a more or less irreversible process like ankylosing in one position a joint which should be flexible. I propose therefore to see what light, if any, can be thrown on some of our assumptions by considering whether a plausible world or a coherent experience might not exist in which they are not fulfilled.

It is usual to begin with time and space. I remember convincing myself of the arbitrary character of Euclid's or any equivalent parallel postulate by imagining myself into a ' Riemann's ' or elliptical space, in which all coplanar lines meet once. I was standing on a transparent plane. I could see it as I looked down. If I looked up I saw the other side of it, and through it the soles of my boots, pointing backwards. By looking round I could see every point on the plane, and most of them from both sides. I soon began to get intuitive proofs of many of the more elementary propositions in that rather bizarre geometry. I therefore ceased to trust ' proofs ' of that type in Euclidian geometry. Of course, any mathematician with a visual imagination can do this, and Einstein has left common-sense space in a badly damaged condition. So we will consider some possibilities about time. Time is more interesting and inaccessible than space because it is given in our inner experience as well as our experience of the world. I am now aware of a 'specious present' of experience about two seconds in length at most, in which I see moving objects and hear sound sequences. I cannot, however, be directly conscious at the same time of a series of events lasting for more than two seconds. A long life consists of about a thousand million specious presents or 'nows.' Of course they overlap, but it is convenient to take them as units. My consciousness at the present moment is in a special relationship to that at other moments in my past. It remembers a few of them, and is influenced by many of them. It has not got this relation to events in my future, or in your past or future. The fact that relations of this type exist determines my personal identity and also my knowledge of time. The perception of change, *e.g.* motion, within a specious present might still exist with a different type of relation between specious presents. There is nothing inconceivable in my looking out of the railway window in 1924 at objects which I am passing, being conscious of the motion, and remembering performing the same journey in 1923 and 1925. In this present world unless gifted with second sight I can only remember the former. If therefore we can imagine a different type of relation between 'nows,' there is no need to postulate a very different content of each from the normal in order to find ourselves in a different world. I begin with two possibilities which are quite probably realized, though not by normal men; namely, that Smith remembers that twenty years ago he was Jones and also

Robinson, while Macgregor and Stuart each remember that twenty years ago they were Johnston.

Either a very large number of animals have no memory whatever, or something of this kind happens. If we divide a flatworm in two, both halves may live happily ever after. If each gets a fair share of the nervous system, presumably they get a certain amount of memory from their common parent. And the converse holds when two protozoa fuse. The case of dissociated personality in men is hardly apposite, as two different personalities rarely if ever seem to be fully conscious at the same time. Human consciousnesses do not usually split or unite in this way because human bodies do not. If, on the other hand, as is very widely supposed, consciousness may continue without a body, I see no reason why such restriction should hold. But I leave it as a problem for a person sincerely desirous of immortality whether he would prefer that 100 years hence fifteen distinct spirits each remembered having been he, or that one spirit remembered being he and also fourteen other people. For clearly if 100 years hence some one remembers having been I, I have not died, even though he is less like me than I am now like myself at four years old. Renan suggested that science would progress so far that our successors would be able to reconstruct the past in complete detail, and finally get their consciousness into a relation of memory with our own, thus achieving the resurrection of the just.

Chains of specious presents constituting spiritual beings like ourselves or those we have so far considered are one-dimensional, and are naturally represented by lines which may fork or unite. They might also form a closed curve, experience repeating itself endlessly after the passage of a certain period. There is also, so far as I can see, nothing inconceivable in two specious presents each remembering the other. Much more interesting, however, is the question of two-dimensional time. There are many possible types of spiritual being enjoying two-dimensional time, just as there are many types of two-dimensional space, but I shall only describe one type. It consists of specious presents like our own, and such that if A remembers B, B never remembers A. On the other hand, B may not remember or be in any way affected by C, and conversely, though both are remembered by A, and can remember D. If we represent instants on a plane, one specious present remembers another only if the latter is neither north nor east of it. Such a being is at any moment aware of a two-dimensional ' creative advance,' which can no more be imagined *sensu stricto* than four-dimensional space. On the other hand, like four-dimensional space, it is easy to reason about it, and one does so whenever one discusses a function of two independent variables. Such a spiritual being would have the great advantage over ourselves of being able to eat his cake, have it, and compare the two experiences. Now, it is clear that as the events which we call the material universe have only one temporal dimension, there is not, so to speak, room for a being of this type among them. On the other hand, there is no reason why my present consciousness should not constitute one edge of a two-dimensional spirit. The chief reason against the indefinite prolongation of the series of specious presents which constitute my mind is not so much that they apparently end abruptly (for they might recommence elsewhere) as that if I survive beyond the age of 60 or so they will show a progressive deterioration. This deterioration is quite obviously related to that of my body. If, on the other hand, the one-dimensional character of time is due to the nature of the ' material ' world rather than that of the mind, our desire for immortality is more likely to be satisfied in some other time-series than in that associated with our bodies.

But let us return to minds of which we know something. How does the world appear to a being with different senses and instincts from our own; and if such beings postulated a reality behind these appearances, what would they regard as real? We will begin with an animal like the dog, which possesses all our senses, although smell is vastly more important for him than for us, and whose instincts are sufficiently like our own to create a bond of real sympathy between us. Of course, we shall have to imagine a dog far more intellectual than any which exists, a dog with a brain as well organized as our own, though organized on rather different lines. And we must give him a language, for even in Laputa thinking without symbols was not very successful.

Now, perhaps man's greatest intellectual achievement is the idea of a thing, by which I mean a portion of experience conceived of as public and ethically neutral. Public because you and Brown experience it as well as I; ethically neutral because it is not good like a man, timid like a rabbit, painful like a headache, appetizing like a taste. We do not regard a pin as being painful in itself or sugar pleasant. We say that the one causes us pain when it enters our foot, the other pleasure when it enters our mouth. This is because we [1] regard its size, shape, hardness, and perhaps colour, as the most real things about the pin. Size is, on the whole, an ethically neutral quality, though when very large it inspires awe. Shape is nearly neutral except by association. Sounds are far less so. An ugly discord is far more disturbing than an ugly pattern. But smells are near the other end of the scale. In the language of physiology, smells normally, sights and sounds very rarely, arouse unconditioned reflexes. Men on the whole divide them into good and bad, though from the emotional point of view ambivalent smells, such as that of trimethylamine, are the most interesting. Psycho-analysts, anticipated by Montaigne, have commented sufficiently on the text ' cuiusque stercus sibi bene olet,' though this is truer of man than dog. It is impossible to observe a dog sniffing about without concluding that much at least of what he smells has a direct emotional effect on him. Clearly association plays a part, but certain smells, as such, appear to waken, for example, hunting and social instincts. Now, an object's most interesting quality to a dog, and that about which he could say most, is its smell. But just for this reason most objects inevitably and spontaneously call up their appropriate emotion, and would always appear with a tertiary quality. The dog's world must be much more like that of the poet Wordsworth, or of a primitive animistic savage, than

[1] I am talking of the average man, not the physicist or metaphysician.

our own. He would be as unable to think impartially and coldly about many ordinary objects as we about our neighbours. I doubt if a dog would ever arrive at our idea of a *thing*, at least for objects with interesting smells. He would find our religion much more intelligible than our science. For one of the essential features of religion is the investing of certain objects (or in later forms the whole universe) with qualities which the scientific point of view ignores. Books, buildings, foods, drinks, rivers, or stones are holy. They not only arouse a special emotion in the believer, but are thought to have these qualities apart from their relation to him, often as the result of ceremonies performed on them. This is just what the dog must feel. If dogs had a religion they would certainly flood their holy buildings with that ' doggy ' smell which is the material basis of their herd instinct. But the affective quality of smells would make it very hard to compare them with regard to their non-affective qualities, a procedure which is the basis of science. Although it is true that many of man's intellectual and moral ideas originated in the emotional atmosphere associated with magic and religion, and even within religious organizations, yet where they have developed it has always been in an environment where clear thinking and frank criticism were possible.

But let us suppose our dog to have overcome these difficulties to some extent, and to be in a position to classify things according to their smells, as we do according to their sizes when we measure them. He will clearly distinguish degrees of intensity of a given smell. But he will also be able to order smells according to their quality, as we do when we classify musical notes according to their frequency or colours by their degree of saturation. For example, a dog is enormously sensitive to the odours of the volatile fatty acids. Buytendijk found that dogs could distinguish between solutions of one part in a million of caproic and caprylic acids by sniffing at them. He would probably be able to place the acids with an even number of carbon atoms in the order of their molecular weights by their smells, just as a man could place a number of piano wires in the order of their lengths by means of their notes. Having got the smells of a number of objects in their right order, as we do with points in a line, how would he proceed to establish relations between them such as we arrive at when we say that the points A and C are equi-distant from B ? He might do it by a method of mixtures. If smells A and C when mixed are indistinguishable from B, he would be justified in believing in a relation of this kind. But now suppose that smells A, B, C, and D, due to the vapours of the substances a, β, γ, and δ, form a series of this nature ; and it is found that the vapour of a subtance ϵ, which is itself inodorous, gives the smell D when mixed with the vapour of γ. Remember that to a dog a thing's smell is its most real quality. He uses the term ' smellable ' or ' odorous ' to denote ' reality,' just as we use ' tangible ' or ' visible.' I think he will say that the vapour of ϵ has a virtual or imperceptible smell E, continuing the series A, B, C, D. The idea of a virtual smell will become clearer if we consider another process used by our scientific dog, analogous to transposition in acoustics and magnification in optics.

The dog performs some process on himself such as fatigue of the sense of smell, which causes the substance β to have the smell A, γ the smell B, δ the smell C, and ϵ the smell D. (I have shown elsewhere that a process not unlike this is possible in man.) He will clearly say that the operation has made the smell E, which was previously imperceptible, appear to be D. When we look at a previously invisible object with a microscope we use the same argument. 'Here,' we say, 'is a microscope which makes this fly look a thousand times its real size. Hence, corresponding to this other oblong image which I see, there must be a small and intangible object of one-thousandth its size. I will call it a bacillus.' We do not dream of questioning the reality of such invisible and intangible objects, and down to the size of bacteria our assumptions work very well. But we cannot magnify objects much smaller than a wave-length of light; and yet we go on supposing that space has still just the same properties as the space in which we find that the evidences of our vision and touch agree with one another. It is not until we get down to the dimensions of an atom that space and time cease to have the properties familiar to us.

If we are surprised at this we must return to the dog. He was so impressed with the reality of smells that he took every opportunity to postulate smells, even when they were unsmellable. If he had talked of inodorous vapours he would have been nearer the truth. We men are at the moment so impressed with the reality of size, shape, and motion, that we postulate objects which we can either touch or see, but which have size, shape, and motion, but no colour, sound, smell, or taste. If we insist on doing this we find that the objects so postulated have to obey quantum mechanics, which begin where the Red Queen in *Through the Looking-Glass* left off. Similarly, if we insist that the optical and mechanical properties of large or rapidly moving bodies should agree, as do those of the small and slow bodies of ordinary life, we are landed in relativist mechanics, which the Red Queen partially anticipated. We do not yet know what our descendants will regard as more real than objects with definite sizes, shapes, and positions. The theory of relativity suggests that they may think happily about space-time with peculiar local properties, while quantum mechanics might lead one to suppose that they will believe in atoms of action or of angular momentum. But the sceptical dog who doubted the theory that everything had a smell would have equal difficulty in knowing what to put in its place.

But the dog is too near to us. Let us go to the insect. J. S. Huxley, in *Essays of a Biologist*, has elaborated one of their troubles in world-building at some length; but perhaps it is Lord Dunsany, in *The Flight of the Queen*, who has given the most vivid imaginative picture of insect psychology in our language. Our own religious feelings at their most intense are perhaps only a feeble shadow of the normal emotional life of a social insect. We mammals are torn between selfish and social desires. There is very little evidence of any such conflict in the life of a normal worker bee, though it is true that some bees are lazier than others. It is largely out of this conflict that reflective thought has arisen in man. In the Bible the origin and nature of the world are dismissed in the first one-and-a-half chapters. The rest is mainly occupied with the results of the conflict between different human instincts. Even the account of the creation only sets the scene for the origin of that conflict in the garden of Eden. And where afterwards historical events are recorded they are the mere background for the moral conflicts of individuals, and for the story of how the tribal god of Israel, who had personified the consciousness of solidarity in one small tribe, gradually developed into the judge of all the earth, the father who is postulated to explain the brotherhood of man. If man had followed a single set of instincts he would never have come to reflect on moral problems, and it is out of this reflection that the great religious systems at any rate have sprung. Animism and polytheism were succeeded by monotheism in Israel on moral and political grounds. Baal was not rejected because the orderly character of nature suggested a single governor of the universe, but because the Lord was regarded as a jealous god. The unity of God is rather the sign of man's attempt to unify his own moral life by following one law, than an explanation of the reign of law in the world. It is true that the Greeks were arriving at the idea of the world unity from a somewhat different direction. But their conception of fate or $\dot{a}\nu\dot{a}\gamma\kappa\eta$ (necessity) seems also to have risen historically from a consideration of moral rather than physical problems. And a moral problem can only arise from the conflict of instincts.

Out of the mythology of the early religious systems arose the attempts to explain nature. The views of the Ionian philosophers belong to the same order of thought as those of the mythologists, and in the systems of such thinkers as Plato and Pythagoras it is impossible to separate theological and physical hypotheses. But the bee has no need for a religious system. Its behaviour on most occasions is prescribed for it by instinct. If it could tell us of the world I think it would speak of a system of duties rather than a system of things. Its language would be one of verbs rather than nouns. The reality of a flower would be the sucking of honey or the gathering of pollen, rather than the flower's form, colour, or odour. It is only because for us most things may have more purposes than one that we do not think of them in this way. Things are specially simplified for the bee because in any situation all the workers of a given age have the same duty, except in so far as they specialize on one flower rather than another. The average worker would not aspire to imagine herself in the position of a queen, while she would regard drones as duty-blind and execrable creatures.

An expression in human terms of the superior reality of duties, as compared with things or even souls, may be found in the Mohammedan (or rather Sunni) dogma of the uncreated Koran. Before there were men and women their reciprocal rights and duties existed, we are told, and the prohibition of wine-bibbing preceded the creation of the grape.

Within our own species those who are conscientiously and successfully engaged in simple and primitive forms of activity would seem to come nearest to living in such a world as the bee. A successful and hard-working mother of a large family is apt, even to an irritating extent, to know the right thing to do in every circumstance. As far as she herself is concerned she very often does; but she tends to be equally certain of the duty of other people, and hence to a certain narrowness in her moral ideas. Primitive men generally seem to know the right thing to do in most of the circumstances of their normal life. Detailed moral tradition occupies the place in their lives that instinct does in that of an animal. And as we discard these dummy instincts we feel a moral nakedness, as it were, which we try to cover in various ways, often strange and inadequate.

I do not see why we should deny the bee the reality of her duty world. Duties are, I suspect, as real as material things, which is not perhaps saying very much. Unfortunately, such an admission is generally taken to imply a belief in the infallibility of the utterances of the moral consciousness. Conscience appears to me to be no more infallible than perception. I see and feel a lump of iron. It appears quite solid, and its parts seem to have no motion relative to one another. I investigate its properties and find that it consists almost wholly of ' empty space,' with a number of tiny particles moving about in it at enormous speeds. Still, there is some meaning in what I perceived, and my perception of the solidity of the metal can be interpreted in terms which are nearer to the truth. Similarly, I perceive a duty, say to aid my distressed neighbour by giving him a new pair of boots. Very likely my duty in detail is just as different as the real iron from my perception of it. I ought perhaps to leave my neighbour with holes in his boots, and give the price of a new pair to the charity organization society or the communist party. Yet that is not to say that there is no reality corresponding to what at first sight I regard as a solid lump of metal or a duty to clothe my neighbour.   But one may fall into just as great errors by taking the one too seriously as the other.   In the long run we may welcome these difficulties because they make us think, but there are times when I at least am disposed to envy the bee, which has but little occasion for this kind of thought. We shall probably in time reduce duty to something else, as we have reduced matter to electricity, but that will not explain it away.

We can perhaps obtain some notion of the contingency of our ethical and aesthetical values by imagining the condition of affairs if the human race, like many animals, possessed an annual breeding season of short duration.   Every year, in the course of a few weeks, we should undergo the profound changes in almost every department of our own mental life which are actually spread over several years of adolescence.   Not only would the sexual instinct awake, but our tastes in art, literature, clothing, politics, and religion would suddenly alter.   As we impose our adult tastes in these matters on children, we can only dimly guess at the values of a humanity without sexual instincts.

No doubt the sexes would be segregated during the breeding season as they are throughout the year in Mohammedan countries.   Love, which is a synthesis of sexual passion with friendship arising out of common interests, would be almost impossible, and human life would be a poorer thing in many ways.   But in particular it is hard to see how any stable system of moral, political, aesthetic, or religious ideas could come into being.   And even if the absolute character of the ideas and values appertaining to such branches of human

activity be denied, some form of intellectual construction is almost undeniably preferable to raw emotion
as a basis for behaviour in these spheres. An animal
with a breeding season would find little permanent but
material objects, and its philosophy would probably
be a crude materialism, its conduct regulated by a
system of harsh and arbitrary laws rather than by any
internal criterion.

Clearly sexual activity outside the breeding season
would be treated as we treat incest or homosexuality.
One need not attempt to picture in any great detail the
fate of an Australian who reached England in autumn
and behaved as Australians did in spring. What is
important to realize is the fact that we can know an
orderly world only because the waking activities of our
mind are fairly similar from one day to another, and
we have agreed to lay little stress upon our dream life.

In the social animals there is at least some chance of
a thought-provoking conflict between social and in-
dividualistic instincts; but in a non-social animal this
is not so, as Trotter in *Instincts of the Herd in Peace and
War* has pointed out, and its world must be still
farther removed from ours; so that any picture one can
draw of it will be more frankly fabulous than those so
far attempted. Let us try to imagine the world of a
sessile and barely social animal endowed with sense
organs. We will allow it some tentacles like a sea-
anemone, or jointed appendages like a barnacle or
crinoid, some eyes like a scallop, and of course organs of
smell or taste. Naturally its most improbable endowment is a brain, for brains are only of value to mobile
animals which have a reasonable number of choices of

action before them. Hence in reality, since a brainless animal can hardly be aware of the world, the number of world-views possible to organisms on this planet is limited to a small fraction of those possible in the abstract. But a brief sketch of the world as it might appear under barely realizable or unrealizable circumstances may be as valuable as the mathematical study of the less realizable types of geometry.

We return then to our philosophical barnacle. It is true that in its youth it was a free-swimming microscopic larva, but it will probably no more remember this than we can remember the time when we were sessile and absorbed our nourishment through a stalk. When we know more about the factors controlling the growth of our own nervous system, we may be in a position to cause it to develop sufficiently before birth for babies to be born with a fully fledged consciousness, and carry over into separate life some memory of their pre-natal state. But we will spare our barnacle this complication. By the mobility of its arms and stalk it can explore a sharply limited volume of space. Beyond that it can see, but it will have little more idea of distance than our unaided senses give us of the distances of the heavenly bodies. It will have a notion of direction, though even that will be as crude as our own localization of internal sensations without the data derived from exploration with our hands. With the perfection of local anaesthesia many of our descendants will probably be familiar with the feel and look of their own internal organs. They will take advantage of surgical operations to know themselves in this sense.

The barnacle, then, finds as great a difficulty in

unifying its visual and tactile space as an astronomer in calculating the distances of the stars. In fact, the average sensual barnacle regards the attempt to do so as ludicrous and presumptuous. ' The world,' it says, ' is what we can sweep with our arms. Things come into it, and my visions are of some use to me in telling me of things that will come into being in it, but they are notoriously deceptive. I know that when a vision becomes very large it is time for me to shut my shell, though sometimes even a very large vision does not portend any real event. But that rule of conduct was revealed to us by the Great Barnacle ages ago, and was not discovered by the philosophers. I also know that when I have a vision in a certain direction, a real thing will sometimes come into being within range of my fourth left arm, and so on. But it seldom does, or I should be fatter than I am ! I do not think that we are helped in any way by calling visions " near " if they precede the advent of a real thing, and " far " if they do not. Visions are visions and realities are realities, and no good will come of mixing them up. A philosopher on the next rock was telling his neighbours that a large vision was " far " and not dangerous, when a thing came into being and nipped six of his arms off. His neighbours had all shut up, and he got little pity from them ! ' Nevertheless, a number of earnest barnacles have formed a society for the investigation of visions. They find that though they generally agree in seeing a vision at the same time, they often differ about its shape and direction. The sceptics say that this proves that visions are nonsense. The members of the S.V.R. (Society for Visionary Research) have recorded many series of partial correspondences between visions of different individuals, and believe that they are on the track of some law governing them. Unfortunately, they are handicapped by two causes besides scepticism. A number of barnacles hold that after a barnacle has died it becomes a vision; while others, inspired by a love of gain or notoriety, make claims which can hardly be substantiated to seeing visions. So on the whole the sensible barnacle considers that there is nothing real out of range of his own or his neighbour's arms. Some of them would qualify this by a statement that bad barnacles when they die go to a rock where it is always low tide, while the virtuous are planted near the opening of an immense sewer, where food is carried to their mouths without any effort on their part. But the idea that so fixed and respectable an animal should be transformed into a vision, which is not only unsubstantial, but mobile, they regard as merely disgusting.

Such is the intellectual condition of the English barnacle. Those living on the coast of Madagascar (*Lepas sapiens*) have worked out the theory of the parallax of visions. They have shown that if the direction of a vision from any two barnacles is known, that from a third can be calculated, and they have developed a satisfactory mathematical theory of visions of which I am privileged to give a brief sketch. A vision as perceived exists in two-dimensional space, the co-ordinates generally employed being front and back, right and left. A given barnacle fixes the position of a vision in its own visual space. It then receives messages from its friends as to where they see the corresponding vision, and by a rapid calculation (or

by means of tables) evaluates a third or imaginary
co-ordinate for the vision. This is often called the
distance. When it becomes zero (or very small) the
vision is associated with a real object. So much is
conceded by all students of the higher mathematics.
But some have gone so far as to suggest that the
imaginary co-ordinate has the same reality as the per-
ceptible ones, and that, in fact, visions exist in a three-
dimensional world. It is admitted that such a world
cannot be imagined, and no barnacle takes it seriously,
though a few of them pretend to. It certainly leads
to somewhat incredible consequences. For example,
when the distance became negative the vision would be
located on the other side of the surface of the rock.
This is an obvious contradiction in terms, for it is well
known that space cannot exist in the absence of water,
and the surface of the rock is the end of space. (I owe
this valuable idea to Mr. George Bernard Shaw, who
in the course of a conversation doubted whether the sun
was more than a few hundred miles away. The so-
called interstellar space, he stated under cross-examina-
tion, has not the properties of ordinary space. It will
not conduct sound, nor can a human being move
through it. It is therefore illegitimate to measure it
in miles.)

Man is after all only a little freer than a barnacle.
Our bodily and mental activities are fairly rigidly confined to those which have had survival value to our
ancestors during the last few million generations. Our
own appraisement of these activities is dictated to some
extent by other considerations than their survival value,
but their nature is limited by our past. We have

learned to think on two different lines—one which
enables us to deal with situations in which we find ourselves in relation to our fellow-men, another for
corresponding situations with regard to inanimate
objects. We are pretty nearly incapable of any other
types of thought. And so we regard an electron as a
thing, and God as a person,[1] and are surprised to find
ourselves entangled in quantum mechanics and the
Athanasian Creed. We are just getting at the rudiments
of other ways of thinking. A few mystics manage to
conceive of God as such, and not as a person or a substance. They have no grammar or even vocabulary
to express their experience, and are generally regarded
as talking nonsense, as indeed they often do. We
biologists, or some of us, are managing to think about
an organism neither as a mere physico-chemical system,
nor as something directed by a mind. We also tend to
contradict ourselves when we try to put our ideas into
words. On the other hand, our way of thinking has led
some of us to a very shrewd idea of how an organism
will behave in given circumstances, and to making
experiments which throw a good deal of light on the
nature of an organism. But I do not feel that any of us
know enough about the possible kinds of being and
thought, to make it worth while taking any of our metaphysical systems very much more seriously than those
at which a thinking barnacle might arrive. Such
systems seem to be helps to the imagination rather than
accounts of reality. Yet it is of fundamental import-

[1] It is only fair to Christianity to point out that belief in a
personal God is heretical, the Almighty being a Trinity, and in
some ways more like the perfect state than the perfect person.

ance that metaphysical speculation should continue. The only alternative to this appears to be the adoption of some rather crude metaphysical system, such as Thomism or materialism, and regarding it as common sense.

But let us return to our fables. So far we have considered animals with an idea of space comparable with ours, an idea derived from vision or from that combination of reaching and locomotion which is possible in the blind,[1] and leads to an idea of space not very different from our own. But there are senses of great delicacy and scope which no more than hint at space. Our sense of hearing is one, and with its aid a system of music has been built up so vast and detailed as almost to constitute a world. Unfortunately, it is not the world in which our bodies live, and hence the bodies of musicians tend to starve in garrets and to place their associated minds in even less dignified situations. But how would the real world appear to a being with a complete series of senses which perceived periodic disturbances as qualitatively different, like our own senses of tone and colour ? We will give it a range of seventy octaves, which would make it aware of the whole range of vibrations from one per second up to the unimaginably but not incalculably high frequency of γ rays from radio-active elements. And within each octave we will endow it with what we possess in our tone sense but not in our colour sense, a capacity for analysing mixed vibrations into their components, as a spectroscope does. Like a musician, too, it will be able to place the various types of radiation in a scale like

[1] Villey, *Le Monde des Aveugles.*

that of musical notes.   It is a curious fact that we men can place musical notes in their natural order by intuition, while it required the genius of Newton to do the same for colour.   What is more, we know that an octave in one part of the scale is equivalent to an octave in another, and hence our musical scale is quantitative. Indeed, in the chromatic scale the notes are so arranged that to each interval between two of them corresponds the same difference in the logarithms of their frequencies.   The piano keyboard is really a rather inaccurate table of logarithms, a fact which I believe is equally ignored in the teaching of mathematics and of music.

But to return to our hypothetical organism, one can point at once to some of its powers.   It could distinguish any chemical substance from any other by the difference in their capacities for absorbing radiation. We men can distinguish a few by their capacities for absorbing visible rays, which give them their different colours, though our colour sense is so inadequate that we have to fall back on the spectroscope.   Our organism could also tell the temperature of any object by analysing the radiation from it.   So that from the qualitative point of view it would know far more than we about objects within the range of its senses.   But it would only arrive at their shapes, sizes, positions, and motions by a most complicated process of deductions, the reverse of the process which we use to discover the nature of the periodic disturbances in molecules.   With no other sense than that described above, its task of world-making would be more hopeless than that of a blind and deaf man.   We must allow it a rudimentary appreciation of space and motion, just as we have a rudimentary appreciation of radiation in our colour sense. It must have at least one movable organ, and be conscious of moving it. It will, however, take colours, if we may so describe the data of its vibration sense, for granted, and build up everything else on this basis. It will, of course, analyse all kinds of motion into periodic components, just as we analyse them into movements in various directions. But it will also, at first, at any rate, regard matter as merely a kind of vibration, or colour, and only very gradually, if ever, reach a point of view like our own.

Now, the oddest thing about its endeavours is that they are of the greatest importance for physicists to-day, and probably of the greatest practical importance to our grandchildren. A century ago physicists began to give up the corpuscular theory of light, which had satisfied them for two thousand years, in favour of a wave theory. Among the practical consequences flowing from this theory were wireless telegraphy and telephony. And in the last two years a much more surprising step has been taken. The wave theory of matter, enunciated by the Duc de Broglie, and developed by Schrödinger, has already rendered the mechanics of the atom relatively intelligible. It has further enabled mathematical physicists to predict several extremely surprising results which have been verified. In consequence some of the ablest men in the world are at present in the position of the mythical creatures which I have tried to describe. They take as their data the frequencies of the radiation emitted or absorbed by various kinds of matter, and very naturally come to regard the matter itself as merely a special type of undulatory disturbance.

So far as an outsider can judge, even Schrödinger's world, fantastic as it is, contains many relics of ordinary thought which the creatures that I have imagined would hardly have taken for granted. However, Heisenberg and Born in Germany, and Dirac in Cambridge, are busily clearing away these vestiges of common sense. In the world of their imagining even the ordinary rules of arithmetic no longer hold good. The attempt to build up a world-view from the end which common sense regards as wrong, is, at any rate, being made, and with very fair success. I suspect that it is of far greater importance for metaphysics than the entire efforts of the philosophers who, from Kant onwards, have attempted to build on the ground cleared by Hume. If it were successful it might lead to philosophical systems in which the real elements in the external world were the secondary qualities of colour, tone, and so forth, rather than the primary qualities of the materialist's world. One may perhaps speculate that in colour vision we have a real perception of light quanta, though the analogy with hearing renders such a theory dubious.

A natural philosophy of such a kind would be a step in the direction of idealism. The idealists have held that the spiritual alone is the real. They have failed to account in detail for the phenomenal world on this basis, the most magnificent of such failures being Hegel's. (I call to mind an admirable picture by a deceased friend entitled 'An Hegelian setting the Dialectic in motion.' A small, bald, and myopic philosopher is turning the handle of a vast and complicated machine, fed from sacks labelled ' Ideen.' It has numerous doors at different levels. That which happens to be open is disgorging rabbits of various colours. That below would have presumably produced plants, that above ' subjective minds.') But the failure of these philosophers in detail does not prove that they were not correct in a general way. Secondary qualities, such as colour, are generally regarded as having less claim to independence of the mind than primary qualities, such as size and shape, and a working theory of the universe which started from them would certainly be a long way nearer to idealism than is present-day science. If, as Leibniz held, the universe consists wholly of minds, the transition to such a physics would only be a step in the right direction, but possibly subsequent steps might be easier. Perhaps an understanding of the psychology of social insects might help us to make them.

I greatly doubt if they will be made by professional philosophers. And though to-day the theoretical physicist is and ought to be the principal type of world-builder, the biologist will one day come to his own in this respect. And one day man will be able to do in reality what in this essay I have done in jest, namely, to look at existence from the point of view of non-human minds. Bergson has of course made this attempt, but not, as it seems to me, very successfully. Success is, indeed, impossible in view of our present ignorance of animal psychology, and that is why a purely speculative essay like the present can claim some degree of justification at this moment. Our only hope of understanding the universe is to look at it from as many different points of view as possible. This is one of the reasons why the data of the mystical consciousness can usefully supplement those of the mind in its normal state. Now, my own suspicion is that the universe is not only queerer than we suppose, but queerer than we *can* suppose. I have read and heard many attempts at a systematic account of it, from materialism and theosophy to the Christian system or that of Kant, and I have always felt that they were much too simple. I suspect that there are more things in heaven and earth than are dreamed of, or can be dreamed of, in any philosophy. That is the reason why I have no philosophy myself, and must be my excuse for dreaming.

THE LAST JUDGMENT

'Denique montibus altior omnibus ultimus ignis
Surget, inertibus ima tenentibus, astra benignis,
Flammaque libera surget ad aëra, surget ad astra,
Diruet atria, moenia, regna, suburbia, castra.'

BERNARD OF CLUNY,
De Contemptu mundi, Lib. I.

THE star on which we live had a beginning and will doubtless have an end. A great many people have predicted that end, with varying degrees of picturesqueness. The Christian account contains much that is admirable, but suffers from two cardinal defects. In the first place, it is written from the point of view of the angels and a small minority of the human race. The impartial historian of the future could legitimately demand a view of the *communiqués* of the Beast of the Book of Revelation and his adherents. For, after all, the Beast and his false prophet could work miracles of a kind, and were admittedly able propagandists. So perhaps 'Another air raid on Babylon beaten off. Seventeen archangels brought down in flames' might record some of the earlier stages in the war, while 'More enemy atrocities. Prophet cast into burning sulphur' would chronicle the peace terms.

But the more serious objection is perhaps to the scale of magnitudes employed. The misbehaviours of the human race might induce their creator to wipe out their planet, but hardly the entire stellar system. We may be bad, but I cannot believe that we are as bad as all

that. At worst our earth is only a very small septic
area in the universe, which could be sterilized without
very great trouble, and conceivably is not even worth
sterilizing.

I prefer *Ragnarok*, the Doom of the Reigners, which
closes the present chapter in world history according
to Norse mythology. Here mankind perish as an
episode in a vaster conflict. It is true that they mis-
behaved first.

> ' Hart es i heimi, hordomr mikkil,
> Skeggi-aold, skalm-aold, skildir klofnir,
> Vind-aold, varg-aold, aðr vaerold steipisk [1]

> (' Hard upon earth then, many a whoredom,
> Sword-age, axe-age, shields are cloven,
> Wind-age, wolf-age, ere world perish '),

says the Norse Sybil in the *Volospa*. But human
events are a symptom rather than a cause. The gods
are to be destroyed by the powers of darkness. Fenri,
the wolf, will eat Odin, and actually get the world
between his teeth, though he will fail to swallow it.
There is a happy ending, probably due to Christian
influence. Balder returns from the dead, and rules
over the descendants of two survivors of the human
race. But one episode is of considerable interest. In
the middle of the fight the sun becomes a mother, and
both she and her daughter survive it. In Scandinavia,
of course, the sun, who is kindly but rather ineffective,
is a female, a conception impossible to the inhabitants
of hotter climates.

Now, fission is one of the vices to which suns are

[1] The letter ð was pronounced like th in then, þ like th in thin.

subject. Indeed, something like half the 'fixed' stars known to us are double or multiple. Apparently the reason for splitting is as follows :—A star always has a certain amount of angular momentum, or spin, due to its rotation on its axis. As it loses heat it gets smaller, but keeps the same amount of spin. So it has to go round faster, and finally splits in two, like a bursting flywheel, owing to its excessive speed. The sun certainly does not seem likely to do this, for it turns round its axis only once in about four weeks ; whereas in order to split, it would have to do so once in less than an hour. But we can see only its outside, and last year Dr. Jeans, the president of the Royal Astronomical Society, suggested that the sun's inside might be rotating much faster, and that no one could say that it would not divide to-morrow. Naturally, such an event is rather unlikely. The sun has gone on for several thousand million years without doing so. But it is apparently possible.

The results for the earth would be disastrous. Even if the sun's heat did not increase so greatly as to roast mankind forthwith, the earth would cease to revolve in a definite orbit, and year by year would approach the pair of suns nearer at one season, retreat from them further at another, while they themselves would gradually separate, and therefore approach nearer to the earth. Long before a collision occurred we should have come so close to one of them that, under the radiation from a sun covering perhaps a tenth of the sky, the sea would have boiled over and mankind perished.

The sun might perhaps do several other things. It

might cool down, and a generation ago it seemed very plausible that it would do so within a few million years. But as we now know that for the thousand million years or so since the first ice-age recorded by geology it has not got much cooler, there is no reason to suppose that it will begin to do so for a very long time indeed. Modern physics suggests, indeed, that it will shine for at least a million million years. But before that time comes, something very strange, as we shall presently see, will have happened to our own planet.

Stars occasionally burst, expanding enormously, giving out a vast amount of heat, and then dying down again. No one knows why this occurs, but it does seem to happen to stars not at all unlike the sun. If it happened to the sun, the earth would stand as much chance of survival as a butterfly in a furnace. But these explosions are also rare. No star at all near to us has exploded during human history. If Sirius, let us say, exploded in this manner, he would send nearly as much light to the earth as does the moon, and would be visible by day. We cannot say whether this kind of ending for our world is likely or not until we know more as to why it happens to other suns than our own.

Others have suggested a comet or some stray heavenly body as a destroyer. Against this we have the fact that on all the continents nothing more than a few miles in diameter can have fallen in the last few hundred million years. The great meteor imbedded in the desert in Arizona may have formed part of a comet, and some of the scars on the moon may be due to collisions with wandering matter. But the improbability of a collision which would desolate any large part of the earth's surface is enormous, even though the Arizona meteorite would have made a considerable mess of London or New York. It has been suggested that a heavy body passing near the earth might drag it out of its orbit. The orderly and nearly circular character of the orbits of all the planets round the sun shows that they have not been greatly perturbed for a very long time, and probably since their formation. One cannot say that they will never be so perturbed, but one can assert that the odds against any such event in the next million years are more than a thousand to one.

All the possibilities that I have catalogued are essentially accidents. Some of them may happen, just as I may be killed in a railway accident ; but just as my body will not go on working for ever, apart from any accidents, so the earth carries with it through space what will certainly alter its conditions profoundly, and very possibly destroy it as an abode of life. I refer to the moon.

Our Scandinavian ancestors did not neglect our satellite in their account of the twilight of the gods.

‘ Austr byr in aldnar i Iarnviði
Ok foeðir þar Fenris kindir,
Verðr af þeim aollom einar nokkar
Tungls tiugari, i trollz hami.’

(‘ Eastward in Ironwood sits the old witch
And breeds Fenri's children,
Of them all one shall be born
Shaped like an ogre, who shall pitch the moon down.’)

Now, here the sybil who described the future to Odin was substantially in agreement with modern astronomy.

The moon will one day approach the earth so close as to be broken up, and very possibly to destroy the earth's surface features. Certain Mohammedan theologians have interpreted the first verse of the Sura called ' The Moon,' ' The hour is come and the moon is split,' as referring to the end of the world. But outside Scandinavia the prophets of doom have generally described the stars as falling out of heaven, which is an impossibility, for the same reason that a million elephants cannot fall on one fly. They are too large.

In what follows I shall attempt to describe the most probable end of our planet as it might appear to spectators on another. I have been compelled to place the catastrophe within a period of the future accessible to my imagination. For I can imagine what the human race will be like in forty million years, since forty million years ago our ancestors were certainly mammals, and probably quite definitely recognizable as monkeys. But I cannot throw my imagination forward for ten times that period. Four hundred million years ago our ancestors were fish of a very primitive type. I cannot imagine a corresponding change in our descendants.

So I have suggested the only means which, so far as I can see, would be able to speed up the catastrophe. The account given here will be broadcast to infants on the planet Venus some forty million years hence. It has been rendered very freely into English, as many of the elementary ideas of our descendants will be beyond our grasp :—

' It is now certain that human life on the earth's surface is extinct, and quite probable that no living thing whatever remains there. The following is a brief record of the events which led up to the destruction of the ancient home of our species.

'Eighteen hundred and seventy-four million years ago the sun passed very close to the giant star 318.47.19543. The tidal wave raised by it in our sun broke into an incandescent spray. The drops of this spray formed the planets, of all of which the earth rotated by far the most rapidly. The earth's year was then only very slightly shorter than now; but there were 1800 days in it, each lasting only a fifth of the time taken by a day when men appeared on earth. The liquid earth spun round for a few years as a spheroid greatly expanded at the equator and flattened at the poles by its excessive rotation. Then the tidal waves raised in it by the sun became larger and larger. Finally the crest of one of these waves flew off as the moon. At first the moon was very close to the earth, and the month was only a little longer than the day.

'As the moon raised large tides in the still liquid earth the latter was slowed down by their braking action, for all the work of raising the tides is done at the expense of the earth's rotation. But by acting as a brake on the earth, the moon was pushed forward along its course, as any brake is pushed by the wheel that it slows down. As it acquired more speed it rose gradually farther and farther away from the earth, which had now a solid crust, and the month, like the day, became longer. When life began on the earth the moon was already distant, and during the sixteen hundred million years before man appeared it had only moved away to a moderate degree farther.

' When these distances were first measured by men the moon revolved in twenty-nine days, and the braking action of the tides amounted to twenty thousand million horse-power on the average. It is said that the effect of tidal friction in slowing down the earth's rotation, and therefore lengthening the day, was first discovered by George Darwin, a son of Charles Darwin, who gave the earliest satisfactory account of evolution. However, there is reason to believe that both these personages are among the mythical culture-heroes of early human history, like Moses, Lao-Tze, Jesus, and Newton.

' At this time the effect of tidal friction was to make each century, measured in days, just under a second shorter than the last. The friction occurred mainly in the Bering Sea between northern Asia and America. As soon as the use of heat engines was discovered, man began to oxidize the fossil vegetables to be found under the earth's surface. After a few centuries they gave out, and other sources of energy were employed. The power available from fresh water was small, from winds intermittent, and that from the sun's heat only available with ease in the tropics. The tides were therefore employed, and gradually became the main source of energy. The invention of synthetic food led to a great increase in the world's population, and after the federation of the world it settled down at about twelve thousand million. As tide engines were developed, an ever-increasing use was made of their power; and before the human race had been in existence for a million years, the tide-power utilized aggregated a million million horse-power. The braking action of the tides was increased fiftyfold, and the day began to lengthen appreciably.

' At its natural rate of slowing fifty thousand million years would have elapsed before the day became as long as the month, but it was characteristic of the dwellers on earth that they never looked more than a million years ahead, and the amount of energy available was ridiculously squandered. By the year five million the human race had reached equilibrium ; it was perfectly adjusted to its environment, the life of the individual was about three thousand years ; and the individuals were " happy," that is to say, they lived in accordance with instincts which were gratified. The tidal energy available was now fifty million million horse-power. Large parts of the planet were artificially heated. The continents were remodelled, but human effort was chiefly devoted to the development of personal relationships and to art and music, that is to say, the production of objects, sounds, and patterns of events gratifying to the individual.

' Human evolution had ceased. Natural selection had been abolished, and the slow changes due to other causes were traced to their sources and prevented before very great effects had been produced. It is true that some organs found in primitive man, such as the teeth (hard, bone-like structures in the mouth), had disappeared. But largely on aesthetic grounds the human form was not allowed to vary greatly. The instinctive and traditional preferences of the individual, which were still allowed to influence mating, caused a certain standard body form to be preserved. The almost complete abolition of the pain sense which was carried out before the year five million was the most striking piece of artificial evolution accomplished. For us, who do not regard the individual as an end in itself, the value of this step is questionable.

' Scientific discovery was largely a thing of the past, and men of a scientific bent devoted themselves to the more intricate problems of mathematics, organic chemistry, or the biology of animals and plants, with little or no regard for practical results. Science and art were blended in the practice of horticulture, and the effort expended on the evolution of beautiful flowers would have served to alter the human race profoundly. But evolution is a process more pleasant to direct than to undergo.

' By the year eight million the length of the day had doubled, the moon's distance had increased by twenty per cent., and the month was a third longer than it had been when first measured. It was realized that the earth's rotation would now diminish rapidly, and a few men began to look ahead, and to suggest the colonization of other planets. The older expeditions had all been failures. The projectiles sent out from the earth had mostly been destroyed by air friction, or by meteorites in interstellar space, and those which had reached the moon intact had generally been smashed by their impact on landing. Two expeditions had landed there with oxygen supplies, successfully mapped the face of it which is turned away from the earth, and signalled their results back. But return was impossible, and their members had died on the moon. The projectiles used in the earlier expeditions were metal cylinders ten metres or less in diameter and fifty or more in length. They were dispatched from vertical metal tubes several kilometres in length, of which the lower part was

imbedded below the earth, while the upper projected.
In order to avoid atmospheric resistance these tubes
were generally built in high mountains, so that when the
projectile emerged it had relatively little air to go
through.   The air in the tube itself was evacuated and
a lid on the top removed as the projectile arrived.   It
was started off by a series of mild explosions which
served to give it a muzzle velocity of about five kilometres per second without causing too great a shock.
When it had left the lower atmosphere it progressed on
the rocket principle, being impelled forward by the
explosion of charges in its tail.   The empty sections of
the tail were also blown backward as required.   It
could be turned from inside by rotating a motor, or
by the crew walking round.

   ' On arriving in the gravitational field of another planet
its fall could be slowed by the discharge downward of
more of its explosive cargo, and to check the final part
of its fall various types of resistance were employed,
and collapsible metal rods were extruded to break the
shock of landing.   Nevertheless, landing was generally
fatal.   As is well known, different principles are now
employed.   In particular, on leaving the atmosphere,
wings of metallic foil of a square kilometre or more in
area are spread out to catch the sun's radiation pressure,
and voyages are thus made on principles analogous to
those employed in the ancient sailing-ships.

   ' The desire for individual happiness, and the fact
that it was achieved on earth, made membership of such
expeditions unpopular.   The volunteers, who were
practically committing suicide, were almost all persons
whose mates had died prematurely, or whose psychology

was for some reason so abnormal as to render them incapable of happiness. An expedition reached Mars successfully in the year 9,723,841, but reported that colonization was impracticable. The species dominant on that planet, which conducts its irrigation, are blind to those radiations which we perceive as light, and probably unaware of the existence of other planets; but they appear to possess senses unlike our own, and were able to annihilate this expedition and the only other which reached Mars successfully.

' Half a million years later the first successful landing was effected on Venus, but its members ultimately perished owing to the unfavourable temperature conditions and the shortage of oxygen in its atmosphere. After this such expeditions became rarer.

' In the year 17,846,151 the tide machines had done the first half of their destructive work. The day and the month were now of the same length. For millions of centuries the moon had always turned the same face to the earth, and now the earth dwellers could only see the moon from one of their hemispheres. It hung permanently in the sky above the remains of the old continent of America. The day now lasted for forty-eight of the old days, so that there were only seven and a half days in the year. As the day lengthened the climate altered enormously. The long nights were intensely cold, and the cold was generally balanced by high temperatures during the day. But there were exceptions.

' Mankind had appeared on earth during a period characterized by high mountains and recurrent ice-ages. Mountain-building had indeed almost ceased, though some ranges and many volcanoes appeared during man's early life. But four ice-ages occurred shortly before history began, and a fifth had devastated parts of the northern continents during the second hundred thousand years of history. The ice had, however, been kept within relatively narrow limits by human endeavour. After the end of this period a huge co-operative effort of the human species had destroyed the remaining ice-fields. About the year 220,000 the ice-cap of Greenland had been gradually melted by the application of tidal energy, and soon after this the Arctic Ocean had become permanently ice-free. Later the Antarctic Continent had been similarly treated. Through most of the first half of human history there was therefore no permanent ice or snow save on a few mountains. The climate throughout the earth became relatively mild and uniform, as it had been through most of the time recorded by geology.

' But as the earth's rotation slowed down, its equator contracted, causing earthquakes and mountain-building on a large scale. A good deal of land emerged from the oceans, especially the central Pacific. And with the lengthening of the nights snow began to be deposited on the uplands in fairly large amounts ; near the poles the sun occasionally failed to melt it during the day, and even where it was melted the subsoil was often permanently frozen. In spite of considerable efforts, ice-fields and giant glaciers had already appeared when the moon ceased to rise and set. Above them permanent anti-cyclones once more produced storms in the temperate regions, and rainless deserts in the tropics.

' The animals and plants only partially adapted themselves to the huge fluctuations of temperature. Prac-

tically all the undomesticated mammals, birds, and
reptiles became extinct.   Many of the smaller plants
went through their whole life-cycle in a day, surviving
only as seeds during the night.   But most of the trees
became extinct except when kept warm artificially.

' The human race somewhat diminished in numbers,
but there was still an immense demand for power for
heating and cooling purposes.   The tides raised by the
sun, although they only occurred fifteen times per year,
were used for these ends, and the day was thus still
further lengthened.

' The moon now began once more to move relative to
the earth, but in the opposite direction, rising in the
west and setting in the east.   Very gradually at first,
but then with ever-increasing speed, it began to approach the earth again, and appear larger.   By the year
25,000,000 it had returned to the distance at which it
was when man had first evolved, and it was realized that
its end, and possibly the earth's, were only a few million
years ahead.   But the vast majority of mankind con-
templated the death of their species with less aversion
than their own, and no effective measures were taken
to forestall the approaching doom.

' For the human race on earth was never greatly influenced by an envisaged future.   After physiology was
discovered primitive men long continued to eat and
drink substances which they knew would shorten and
spoil their lives.   Mineral fuels were also oxidized
without much forethought.   The less pigmented of the
primitive races exhausted the fuel under the continents
on which they lived with such speed that for some
centuries the planet was dominated by the yellow

variety resident in eastern Asia, where mining had developed more slowly; until they too had exhausted their fuel resources. The unpigmented men appear to have foreseen this event, but did little or nothing to prevent it, even when it was clearly only a few generations ahead. Yet they had before them the history of an island in the North Atlantic on which Newton and Darwin are said to have lived, and whose inhabitants were the first to extract mineral fuel and the first to exhaust it, after which they disappeared from the stage of history, although at one time they had controlled large portions of the earth's land surface.

'On the contrary, the earth's inhabitants were often influenced in a curious way by events in the past. The early religions all attached great significance to such occurrences. If our own minds dwell more readily on the future, it is due largely to education and daily propaganda, but partly to the presence in our nuclei of genes such as H 149 and P 783 c, which determine certain features of cerebral organization that had no analogy on earth. For this reason we have undertaken the immense labour necessary to tap the central heat of our planet, rather than diminish its rotation. Even now this process involves a certain annual loss of life, and this was very much greater at first, so much so as to forbid its imitation on the earth, whose inhabitants generally valued their own lives and one another's.

'But if most men failed to look ahead, a minority felt otherwise, and expeditions to Venus became commoner. After 284 consecutive failures a landing was established, and before its members died they were able to furnish the first really precise reports as to conditions on that planet. Owing to the opaque character of our atmosphere, the light signals of the earlier expeditions had been difficult to pick up. Infra-red radiation which can penetrate our clouds was now employed.

' A few hundred thousand of the human race, from some of whom we are descended, determined that though men died, man should live for ever. It was only possible for humanity to establish itself on Venus if it were able to withstand the heat and want of oxygen there prevailing, and this could only be done by a deliberate evolution in that direction first accomplished on earth. Enough was known of the causes responsible for evolution to render the experiment possible. The human material was selected in each generation. All who were not willing were able to resign from participation, and among those whose descendants were destined for the conquest of Venus a tradition and an inheritable psychological disposition grew up such as had not been known on earth for twenty-five million years. The psychological types which had been common among the saints and soldiers of early history were revived. Confronted once more with an ideal as high as that of religion, but more rational, a task as concrete as and infinitely greater than that of the patriot, man became once more capable of self-transcendence. Those members of mankind who were once more evolving were not happy. They were out of harmony with their surroundings. Disease and crime reappeared among them. For disease is only a failure of bodily function to adjust itself to the environment, and crime a similar failure in behaviour. But disease and crime, as much as heroism and martyrdom, are part of the price which must be paid for evolution. The price is paid by the individual, and the gain is to the race. Among ourselves an individual may not consider his own interests a dozen times in his life. To our ancestors, fresh from the pursuit of individual happiness, the price must often have seemed too great, and in every generation many who have now left no descendants refused to pay it.

' The modes of behaviour which our ancestors gradually overcame, and which only recur as the rarest aberrations among ourselves, included not only such self-regarding sentiments as pride and a personal preference concerning mating. They embraced emotions such as pity (an unpleasant feeling aroused by the suffering of other individuals). In a life completely dedicated to membership of a super-organism the one is as superfluous as the other, though altruism found its place in the emotional basis of the far looser type of society prevalent on earth.

' In the course of ten thousand years a race had been evolved capable of life at one-tenth of the oxygen pressure prevalent on earth, and the body temperature had been raised by six degrees. The rise to a still higher temperature, correlated as it was with profound chemical and structural changes in the body, was a much slower process. Projectiles of a far larger size were dispatched to Venus. Of 1734, only 11 made satisfactory landings. The crews of the first two of these ultimately perished ; those of the next eight were our ancestors. The organisms found on Venus were built of molecules which were mostly mirror images of those found in terrestrial bodies. Except as sources of fat they were therefore useless for food, and some of them were a serious menace. The third projectile to arrive included bacteria which had been synthesized on earth to attack l-glucose and certain other components of the organisms on Venus. Ten thousand years of laboratory work had gone to their making. With their aid the previous life on that planet was destroyed, and it became available for the use of man and the sixty terrestrial species which he had brought with him.

' The history of our planet need not be given here. After the immense efforts of the first colonizers, we have settled down as members of a super-organism with no limits to its possible progress. The evolution of the individual has been brought under complete social control, and besides enormously enhanced intellectual powers we possess two new senses. The one enables us to apprehend radiation of wave-lengths between 100 and 1200 metres, and thus places every individual at all moments of life, both asleep and awake, under the influence of the voice of the community. It is difficult to see how else we could have achieved as complete a solidarity as has been possible. We can never close our consciousness to those wave-lengths on which we are told of our nature as components of a super-organism or deity, possibly the only one in space-time, and of its past, present, and future. It appears that on earth the psychological equivalent of what is transmitted on these wave-lengths included the higher forms of art, music, and literature, the individual moral consciousness, and, in the early days of mankind, religion and patriotism. The other wave-lengths inform us of matters which are not the concern of all at all times, and we can shut them out if we so desire.

Their function is not essentially different from that of instrumental radio-communication on earth. The new magnetic sense is of less importance, but is of value in flying and otherwise in view of the very opaque character of our atmosphere. It would have been almost superfluous on earth. We have also recovered the pain sense, which had become vestigial on earth, but is of value for the survival of the individual under adverse circumstances, and hence to the race. So rapid was our evolution that the crew of the last projectile to reach Venus were incapable of fertile unions with our inhabitants, and they were therefore used for experimental purposes.

'During the last few million years the moon approached the earth rather rapidly. When it became clear that the final catastrophe could not be long delayed the use of tide-power was largely discontinued, according to the signals which reached us from the earth, and wind and other sources of power were substituted. But the earth-dwellers were sceptical as to whether the approaching rupture of the moon would entail their destruction, and the spin of the earth-moon system was still used to some extent as a source of power. In the year 36,000,000 the moon was at only a fifth of its distance from the earth when history had begun. It appeared twenty-five times as large as the sun, and raised the sea-level by some 200 metres about four times a year. The effects of the tidal strain raised in it by the earth began to tell. Giant landslips were observed in the lunar mountains, and cracks occasionally opened in its surface. Earthquakes also became rather frequent on the earth.

' Finally the moon began to disintegrate. It was so near to the earth as to cover about a twentieth of the visible heavens when the first fragments of rock actually left its surface. The portion nearest to the earth, already extensively cracked, began to fly away in the form of meteorites up to a kilometre in diameter, which revolved round the earth in independent orbits. For about a thousand years this process continued gradually, and finally ceased to arouse interest on the earth. The end came quite suddenly. It was watched from Venus, but the earlier stages were also signalled from the earth. The depression in the moon's surface facing the earth suddenly opened and emitted a torrent of white-hot lava. As the moon passed round the earth it raised the temperature in the tropics to such an extent that rivers and lakes were dried up and vegetation destroyed.

' The colour changes on earth due to the flowering of the plants which were grown on it for the pleasure of the human race, and which were quite visible from our planet, no longer occurred. Dense clouds were formed and gave some protection to the earth. But above them the sea of flame on the moon increased in magnitude, and erupted in immense filaments under the earth's gravitation. Within three days the satellite had broken up into a ring of white-hot lava and dust. The last message received from the earth stated that the entire human race had retired underground, except on the Antarctic Continent, where however the ice-cap had already melted and the air temperature was 35° C. Within a day from the moon's break-up the first large fragment of it had fallen on the earth. The particles formed from it were continually jostling, and many more were subsequently driven down. Through the clouds of steam and volcanic smoke which shrouded the earth our astronomers could see but little, but later on it became clear that its tropical regions had been buried many kilometres deep under lunar fragments, and the remainder, though some traces of the former continents remain, had been submerged in the boiling ocean. It is not considered possible that any vestige of human life remains, nor can our spectroscopes detect any absorption bands of chlorophyll which would indicate the survival of plants.

' The majority of the lunar matter has formed a ring round the earth, like those of Saturn, but far denser. It is not yet in equilibrium, and fragments will continue to fall on the earth for about another thirty-five thousand years. At the end of that period the earth, which now possesses a belt of enormous mountains in its tropical regions, separated from the poles by two rings of sea, will be ready for recolonization. Preparations are being made for this event. We have largely sorted out the useful elements in the outer five kilometres or so of our planet, and it is proposed, when the earth is reoccupied, to erect artificial mountains on both planets which will extend above the Heaviside layer and enable continuous radio-communication instead of light signals to be used between the two.

' The old human race successfully cultivated individual happiness and has been destroyed by fire from heaven. This is not a cause for great regret, since happiness does not summate. The happiness of ten million individuals is not a millionfold the happiness of ten. But the unanimous co-operation of ten million individuals is something beyond their individual behaviour. It is the life of a super-organism. If, as many of the earth-dwellers hoped, the moon had broken up quietly, their species might have lasted a thousand million years instead of thirty-nine million, but their achievement would have been no greater.

'From the earth it is proposed to colonize Jupiter. It is not certain that the attempt will succeed, for the surface temperature of that planet is 130 degrees C., gravitation is three times as intense as that on Venus, and over twice that on earth, while the atmosphere contains appreciable quantities of thoron, a radio-active gas. The intense gravitation would of course destroy bodies as large as our own, but life on Jupiter will be possible for organisms built on a much smaller scale. A dwarf form of the human race about a tenth of our height, and with short stumpy legs but very thick bones, is therefore being bred. Their internal organs will also be very solidly built. They are selected by spinning them round in centrifuges which supply an artificial gravitational field, and destroy the less suitable members of each generation. Adaptation to such intense cold as that on Jupiter is impracticable, but it is proposed to send projectiles of a kilometre in length, which will contain sufficient stores of energy to last their inhabitants for some centuries, during which they may be able to develop the sources available on that planet. It is hoped that as many as one in a thousand of these projectiles may arrive safely. If Jupiter is successfully occupied the outer planets will then be attempted.

' About 250 million years hence our solar system will pass into a region of space in which stars are far denser than in our present neighbourhood. Although not more than one in ten thousand is likely to possess planets suitable for colonization, it is considered possible that we may pass near enough to one so equipped to allow an attempt at landing. If by that time the entire matter of the planets of our system is under conscious control, the attempt will stand some chance of success. Whereas the best time between the earth and Venus was one-tenth of a terrestrial year, the time taken to reach another stellar system would be measured in hundreds or thousands of years, and only a very few projectiles per million would arrive safely. But in such a case waste of life is as inevitable as in the seeding of a plant or the discharge of spermatozoa or pollen. Moreover, it is possible that under the conditions of life in the outer planets the human brain may alter in such a way as to open up possibilities inconceivable to our own minds. Our galaxy has a probable life of at least eighty million million years. Before that time has elapsed it is our ideal that all the matter in it available for life should be within the power of the heirs of the species whose original home has just been destroyed. If that ideal is even approximately fulfilled, the end of the world which we have just witnessed was an episode of entirely negligible importance. And there are other galaxies.'

EPILOGUE

THERE are certain criteria which every attempt, however fantastic, to forecast the future should satisfy. In the first place, the future will not be as we should wish it. The Pilgrim Fathers were much happier in

England under King James I. than they would be in America under President Coolidge. Most of the great ideals of any given age are ignored by the men of later periods. They only interest posterity in so far as they have been embodied in art or literature. I have pictured a human race on the earth absorbed in the pursuit of individual happiness ; on Venus mere components of a monstrous ant-heap. My own ideal is naturally somewhere in between, and so is that of almost every other human being alive to-day. But I see no reason why my ideals should be realized. In the language of religion, God's ways are not our ways ; in that of science, human ideals are the products of natural processes which do not conform to them.

Secondly, we must use a proper time-scale. The earth has lasted between one and eight thousand million years. Recorded human history is a matter of about six thousand. This period bears the same ratio to the earth's life as does a space of two or three days to the whole of human history. I have no doubt that in reality the future will be vastly more surprising than anything I can imagine. But when we once realize the periods of time which our thought can and should envisage we shall come to see that the use, however haltingly, of our imaginations upon the possibilities of the future is a valuable spiritual exercise.

For one of the essential elements of religion is an emotional attitude towards the universe as a whole. As we come to realize the tiny scale, both temporal and spatial, of the older mythologies, and the unimaginable vastness of the possibilities of time and space, we must attempt to conjecture what purposes may be developed in the universe that we are beginning to apprehend. Our private, national, and even international aims are restricted to a time measured in human life-spans.

> ' And yonder all before us lie
> Deserts of vast eternity.'

If it is true, as the higher religions teach, that the individual can only achieve a good life by conforming to a plan greater than his own, it is our duty to realize the possible magnitude of such a plan, whether it be God's or man's. Only so can we come to see that most good actions merely serve to stave off the constant inroads of chaos on the human race. They are necessary, but not sufficient. They cannot be regarded as active co-operation in the Plan. The man who creates a new idea, whether expressed in language, art, or invention, may at least be co-operating actively. The average man cannot do this, but he must learn that the highest of his duties is to assist those who are creating, and the worst of his sins to hinder them.

I do not see how any one who has accepted the view of the universe presented by astronomy and geology can suppose that its main purpose is the preparation of a certain percentage of human souls for so much of perfection and happiness as is possible for them. This may be one of its purposes, but it can hardly be the most important. Events are taking place ' for other great and glorious ends ' which we can only dimly con-jecture. Professor Alexander, for example, in *Space, Time, and Deity*, suggests that the end towards which ' the whole creation groaneth and travaileth ' is the emergence of a new kind of being which will bear the same relation to mind as do mind to life and life to matter. It is the urge towards this which finds its expression in the higher forms of religion. Without necessarily accepting such a view, one can express some of its implications in a myth. The numerical side of the myth is, I believe, correct, though whether tidal power could be utilized to the extent that I have suggested is a question for the engineers of the future.

Man's little world will end. The human mind can already envisage that end. If humanity can enlarge the scope of its will as it has enlarged the reach of its intellect, it will escape that end. If not, the judgment will have gone out against it, and man and all his works will perish eternally. Either the human race will prove that its destiny is in eternity and infinity, and that the value of the individual is negligible in comparison with that destiny, or the time will come

'When the great markets by the sea shut fast
    All that calm Sunday that goes on and on ;
When even lovers find their peace at last,
    And earth is but a star, that once had shone.'